STO

ACPL ITEM
DISCARDED

623.8
BRINK, RANDALL.
THE SEAGOING COMPUTER

S0-EMD-916

DO NOT REMOVE
CARDS FROM POCKET

2/24/94

ALLEN COUNTY PUBLIC LIBRARY

FORT WAYNE, INDIANA 46802

You may return this book to any agency, branch,
or bookmobile of the Allen County Public Library.

DEMCO

Allen County Public Library
900 Webster Street
PO Box 2270
Fort Wayne, IN 46801-2270

*To that small cadre of pioneers
who saw what a computer could add to
the seagoing life.*

Published by International Marine

10 9 8 7 6 5 4 3 2 1

Copyright © 1994 International Marine, an imprint of TAB Books. TAB Books is a division of McGraw-Hill, Inc.

All rights reserved. The publisher takes no responsibility for the use of any of the materials or methods described in this book, nor for the products thereof. The name "International Marine" and the International Marine logo are trademarks of McGraw-Hill, Inc. Printed in the United States of America.

Library of Congress Cataloging-in-Publication Data
Brink, Randall.
  The seagoing computer / Randall Brink.
    p. cm.
  Includes index.
  ISBN 0-87742-364-4
  1. Boats and boating--Electronic equipment. 2. Microcomputers.
I. Title.
VM325.B75 1993
623.8'504--dc20   93-36350
                                                                  CIP

Questions regarding the content of this book should be addressed to:
International Marine
P.O. Box 220
Camden, ME 04843

Questions regarding the ordering of this book should be addressed to:
TAB Books
A Division of McGraw-Hill, Inc.
Blue Ridge Summit, PA 17294
1-800-233-1128

Printed by R.R. Donnelley, Harrisonburg, VA
Design and Production by Faith Hague
Edited by J.R. Babb, Paula Blanchard, Carolyn Marsh

Contents

Acknowledgments *iv*
Introduction *v*
1 The Role of the Seagoing Computer 1
2 Choosing the Seagoing Computer 19
3 Installing the Seagoing Computer 39
4 Choosing and Linking Computer Peripherals 55
5 Communications Utilities 71
6 Software for the Seagoing Computer 81
7 Maintenance and Troubleshooting of Computers and Peripherals 91
8 Computerizing the Helm and Navigation Tasks 105
9 Using the Seagoing Computer for Navigation 125

Appendices
A GPS Equipment Manufacturers 149
B Telecommunications and On-line Services 151
C Manufacturers and Dealers of Marine-Computer Hardware and Software Systems 157

Glossary of Equipment Terms 171
Index 179

Acknowledgments

I am grateful to all those mariners and others who assisted at the various stages of this book. In particular, I wish to thank Dave Crane of D.F. Crane Associates for supplying a good deal of the technical information, industry news, and photographs.

The very active group of computer users of the Delphi Yacht Club are commended for sharing their experiences, insights, and news of developments.

Thanks also to James Babb at International Marine for acknowledging the need for a book about the role of computers in the marine community, and for spearheading the idea through the complicated maze between concept and publication.

Introduction

One Monday morning a few years ago, I sat in the cockpit of my yacht at work on the first draft of a novel. My old Underwood typewriter was on a fold-down table rigged to the mizzenmast. Around me the air was calm, fragrant with aromas unique to the marine environs and sparkling with the first sunshine. The dock was quiet; most of my fellow boaters were at work ashore. As I sipped from a mug of steaming coffee and looked off across the bay toward the breakers of the Pacific Ocean, it occurred to me how fabulously fortunate I was to be able to work in a setting so conducive to creativity. This was the nautical lifestyle, taken a step beyond to include earning a living.

Later in the day, my work complete, I surveyed the marina to see the usual assortment of people: some stragglers from the weekend, others who had sneaked out early from the office, and a few who, like me, had forsaken the conventional life ashore for the rewards and privations of living aboard. I thought again of my good fortune to have chosen a profession as portable and adaptable as the life of letters. Still, my self-congratulations were tempered with the knowledge that at some point I too would have to troop ashore for a long stint at a shore-bound desk to get my latest manuscript keyboarded, edited, and prepared for submission to my publisher. It was at that moment that I began scheming to find a means of doing it all on board—to eliminate even the relatively small inconvenience of an eventual, obligatory trip ashore.

Way back in 1988, amid the Pleistocene epoch of personal computing, the idea of installing a personal computer aboard a boat was at best a wild idea, fraught with myriad logistical, practical, technical, and financial obstacles. Computers were bulky and expensive; they were not designed for the tight confines of a boat or the rigors of seawater and sea air, not to mention motion, heeling, and fluctuating ship's power. It could be done—there were a few liveaboards even then who had computers—but it wasn't an easy or inexpensive proposition, and the end result was not always reliable.

Determination and ingenuity, more or less required traits of a liveaboard, have solved many a problem in the endless history of technological revolution. So I hauled a big, clunky IBM PC-AT aboard—all one-hundred-odd pounds of it. Included were a suitcase-sized central processing unit (CPU), a cathode-ray tube (CRT) monitor, a keyboard, and a brontosaurus-esque daisy-wheel printer. I managed a jury-rigged setup in my navigation station, leaving enough room to write in the ship's log only if I could reduce the log to fit on 3-by-5 index cards. Forget navigation. (This was long before computers were used much in marine navigation.) The installation cost about $4,500, plus the sacrifice of precious space, but at last I could indulge in the dream of never having to go ashore to work again.

Times have changed.

Today I have a little, 5-pound, 386DX laptop with a built-in 80-megabyte (Mb) hard disk, a full, elegantly backlit VGA screen, and a send/receive fax modem that folds into an 8- by 6- by 1-inch shape and tucks neatly onto a chart shelf. My quiet Hewlett Packard Laserjet printer hides unobtrusively in a locker beneath the table, out of the way. I receive faxes and send memos, letters, documents, and entire manuscripts with awesome ease. The whole installation cost less than $3,000.

Working aboard while still in a marina slip was just the first step. I contemplated a period of cruising, but how would I get work back and forth between boat and publisher on a reasonable schedule?

The answer was the addition of cellular communications, a technology so new that few utilities existed for it until late 1991, when cellular modems that link computer data terminals to single-sideband (SSB) radio transceivers made two-way data communications via satellite a reality.

Once the basic infrastructures and technologies were in place in the computer industry, human enterprise began to invent ways to integrate systems and tasks. Seemingly overnight, the ability to navigate, communicate, obtain weather information, and exchange data between shore stations and seagoing vessels was within the reach of every sailor—from the skipper of a pocket cruiser in Puget Sound to the master of a supertanker in the North Sea.

This book is about what's out there in the way of seagoing computer hardware and software technology, and how the mariner can create a versatile and productive seagoing computer system on board. In other words, it's about how the seagoing computer can help make life aboard your boat all it can be.

Randall Brink
September 1993

1

The Role of the Seagoing Computer

Why a Seagoing Computer?

A scant few years ago, finding a personal computer on board a yacht was relatively rare. The few boats that had them were most likely engaged in some form of commercial marine enterprise or academic research, or outfitted for cruising and living aboard by professional owners who simply could not exist without a personal computer.

With the advent of smaller, tougher personal computers, as well as portable laptop and notebook models, computers are going to sea in far greater numbers. The explosion in computer and space technology, competition among innovators, manufacturers, and sellers, and a

2 The Seagoing Computer

miniaturization of components have combined to drive prices steadily downward for the past five years. The emergence of computer designs built especially for marine applications, and the advent of software programs and electronic on-line telecommunications services targeted at the marine market, have resulted in making the onboard computer as practical and useful as a VHF or SSB radio, Loran, or global positioning system (GPS).

A computer can serve a multitude of purposes afloat. With proper hardware and software, these include word processing and other data-processing tasks; handling complex navigational functions and improving navigational accuracy; sending and receiving messages and materials to and from shore; obtaining up-to-the-minute weather information; keeping track of a yacht's budget, maintenance schedule, and provisions; keeping the ship's log; and even aiding in the design of an owner's next boat. A computer is also invaluable for keeping track of documentation required by the U.S. Coast Guard, insurance registry, customs and immigration services, yacht incorporation, mortgage holders, and others.

For the cruising or liveaboard professional, a computer is also useful for on-line library and database research, transmittal of drafts and finished work, and keeping the books of a professional or business enterprise. Running down a list of professionals who might use computers, one quickly realizes the huge potential for a seagoing computer: writers, lawyers, editors, architects, engineers, legal aides and stenographers, accountants, marine surveyors, investment bankers, customs brokers, stockbrokers, investors, computer programmers, consultants—the list goes on and on. All or a portion of the normal workload in any of these professions could be accomplished aboard a boat; seldom, if ever, would a visit to a conventional office ashore be necessary. The principal obstacle to extended cruising—the need to make a living while at sea—is substantially overcome at last with the aid of a computer.

Even the boatowner who desires only to extend the weekend afloat by an extra day or two can do so. A day spent on board a boat can be as productive as one spent in an office—perhaps even more so, considering the much more pleasant environment! Before you relegate such an idea to the realm of fantasy or dismiss an onboard computer workstation as a prohibitively expensive luxury, take note: If you can afford a boat today, you can probably afford to equip it with the latest in computer technology.

The Seagoing Computer Installation

Let's examine onboard computer installations, beginning with the most basic components and moving toward more sophisticated setups especially useful to liveaboards or extended-cruising boaters. As with most new things aboard a boat, it's best to start with the basics and gradually build a system ideally suited to your needs and interests. For our purposes, I'll assume the boatowner intends to do serious work aboard and wants the latest aids to boat safety, navigation, and management.

Hardware Basics

A basic marine computer installation might consist of a conventional desktop-type personal computer, with its

Figure 1-1. *The personal computer is now a familiar sight in the nav stations of yachts as well as aboard nearly all commercial marine vessels.*

CPU mounted on or beneath a chart table and a keyboard and monitor on top (Figure 1-1). The more space-conscious owner might elect to purchase one of the new laptop or notebook computers that can either be built into the nav-station cabinetwork or kept separate for maximum portability (Figure 1-2).

The advent of these very small, lightweight notebook computers has been a significant development in the computer industry. These units incorporate lightning-fast processors—Intel or derivative 80386 or 80486 chips—with MS-DOS–compatible computers, or various high-speed versions of the Motorola 68000-series processors for the popular Apple Macintosh; high-resolution monitors; and large-capacity hard disks with capacities of 80

Figure 1-2. Laptop, or portable, computers are fast becoming an integral part of the bluewater cruising navigation station. This setup was photographed aboard SeaFinch, *a 47-foot Northwind Cutter out of Portsmouth, England, by owner Kent Pietsch.*

MB or more. By adding an internal fax modem and an equally diminutive printer, a boatowner can achieve a fully integrated system in the space of a briefcase, which has the added advantage of being small enough to be removed from the vessel when not in use. (Note: There are explicit—and ever-changing—definitions pertaining to these small laptop and notebook computers. For our purposes, however, the terms *laptop*, *notebook*, and *portable* will be used interchangeably.)

In the earlier days of personal computing, computers were much more fragile and subject to failure if operated in less than ideal environments. Dust, moisture, movement, and temperature fluctuations were all potentially destructive to the delicate workings of a computer. Neither the original IBM PC (or its hundreds of "clones") nor the Apple Macintosh (Mac) was designed with the rigors of life at sea in mind.

Today's computers are much more sturdily built than their forebears. Cases for monitors and CPUs are built tougher and tighter. The computers employ far fewer moving parts, and the ones that remain, such as hard-disk drives, are far more rugged than they were just a few years ago. A survey of onboard computer users, conducted during research for this book, indicates that today's computers have fewer problems with moisture, corrosion, or movement afloat. This represents a noteworthy improvement over the more primitive installations of even a few years ago. Still, there is sufficient risk in operating the rather vulnerable electronic components in the marine environment to warrant the design, production, and sale of computers intended especially for life at sea.

There are now computers designed specifically for marine use and engineered for the vicissitudes of the marine environment, including the Sea PC, designed and manufactured by D.F. Crane and Associates (Figure 1-3), and the marine rack-mount industrial-strength units built by companies such as Texas Micro (Figure 1-4) and Bison Instruments, Inc. (Figure 1-5). Laptops from Hyundai, Zeos, Dell, Zenith, Toshiba, and others are now being aimed at the marine market as well, and these units, despite their diminutive size and very light weight, boast most if not all of the features of their desktop forebears, such as high-resolution VGA screens,

The Role of the Seagoing Computer 7

Figure 1-3. The SEA PC is a fully weatherized IBM-compatible PC built by D.F. Crane Associates especially for the marine market. (Photo courtesy of D.F. Crane Associates, Inc.)

80- to 120-MB hard disks, fast 20-megahertz (MHz) processor clock speed, and 9,600-baud internal send/receive fax modems. (These strange-sounding and esoteric computer terms are explained fully in the chapters to come. They are also defined in the glossary at the back of this book.)

In planning a fully functional onboard computer installation, most boatowners want word processing and math or spreadsheet functions, facsimile (fax), telecommunications, and navigational functions. This sounds extensive, complicated, and expensive, but it really isn't, since many of these functions use common equipment or compact, relatively inexpensive components. All of these functions can be obtained with an IBM-compatible 386SX computer (or its Macintosh equivalent, such as the low-priced Color Classic) with a 60-MB or larger

8 The Seagoing Computer

Figure 1-4. The Texas Micro PCs, originally built for the adverse conditions of factories and other industrial sites, are also suitable for high-end marine installations where space is not a critical factor. (Photo courtesy of Texas Micro)

hard disk (for storage of your data and software programs), 4 MB of RAM (random-access memory, or that portion of the computer processor's memory where the actual computing functions are carried out), an internal 9,600-baud or greater fax modem, and an inexpensive thermal printer. Add to this hardware installation a package of relevant software, much of which is often "bundled" or included with the computer by the manufacturer or dealer at the time of sale, and the basic onboard system is virtually complete—for around $2,000. Putting the computer to work aboard the boat as an electronic print shop, telegrapher, navigator,

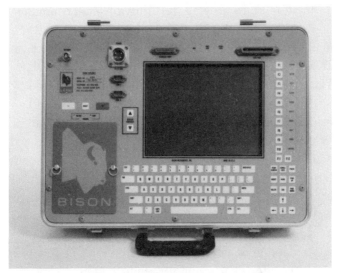

Figure 1-5. Although designed and engineered for such unique tasks as scientific surveys in Arctic weather, the Bison Instruments 486 portable, with its weatherized case and integrated keyboard and screen, would be a contender for a marine installation where the utmost in environment-proofing is desired. (Photo courtesy of Bison Instruments, Inc.)

accountant, secretary, galley slave, harbor pilot, maintenance engineer, and all-around able hand is a matter of software and your own imagination and input.

Seagoing Software

As with most popular interests, yachting has begun to attract its fair share of computer entrepreneurs and programmers. They have produced hardware, software, and networking options for those interested in using computers afloat.

In addition to proving the utility of small computers on board, these individuals have come up with ways to make life aboard easier, better, and more productive. Today there are computer programs such as Bluewater Yacht Manager and Boat Helper, developed and sold by Shelter Island Press in San Diego, California, that can help create provisioning and inventory lists, track routine maintenance, prepare and analyze budgets and financial plans for cruisers, and maintain a boat's legal documentation. Such programs can keep track of navigation charts, crew information, spare parts and equipment, as well as keep the ship's log. There are programs now available through commercial software vendors for navigation and pilotage, including computerized tide tables and navigation charts, and for sending and receiving fax documents (including weatherfax). Advances in cellular and satellite technology now enable the cruiser to access remote computer databases and on-line services through a variety of media, including cellular and satellite uplinks, single-sideband radio "packet" networks, and, of course, conventional telephone lines while ashore.

While the use of personal computers for word processing and bookkeeping on board is not really new, the ability to integrate them fully into the navigation station as a full-featured utility is. The advent of telecommunications accessible on board from anywhere on the globe has made this possible within the past two to three years. Through the use of computers, ocean navigators can not only communicate with anyone anywhere in the world, but can dramatically increase the accuracy of marine navigation—to within 10 yards of any location.

Appendix B includes listings and descriptions of on-line interactive computer services and marine bulletin-board systems (BBSs) that cater to the marine computer user. These services offer a very impressive list of marine computer software programs that increases in size and sophistication every day.

Conventional Software Goes to Sea

Just as I once reveled in the luxury of using my prized WordStar 3.0 word-processing program on my onboard PC, today's leading data-processing software is equally functional when running on the onboard computer.

For devotees of the Apple Macintosh computer, the Apple proprietary operating system, as embodied in the state-of-the-art System 7 environment, offers a graphical user interface (GUI) that allows the user to accomplish all computer tasks on-screen. A pointing device, such as a "mouse," is used to carry out a multitude of computing tasks without the user's ever having to deal with the esoteric language of computer operating systems and programming. The IBM-PC equivalent of the Mac's user-friendly GUI operating system is Microsoft Windows (and, to a far lesser extent, IBM's O/S 2 operating environment).

This book is not about specific computer operating systems, nor is it a definitive guide to the thousands of software applications available to IBM PC and Mac users. For a full description of selected software programs of value to the marine user, particularly those written for marine use, see Chapter 6.

I would like to interject a couple of generalizations here. First, as this is written, Microsoft Windows is the standard operating system in the IBM PC world. The overwhelming majority of professional programmers are writing commercial and shareware programs for the

Windows environment. Second, practically any software program available to PC users under Windows has its equivalent for the Mac. In fairness, the Mac provided the first honest graphical environment for the user, an elegant if obtusely proprietary system that set a high standard for screen displays and user-friendliness, and one that Microsoft Windows was hard-pressed to emulate. Today, the two systems will read each other's software languages and so are virtually interchangeable, at least for our purposes here.

In light of the above, a typical onboard computer installation should include, as a basic software engine, Microsoft MS-DOS Version 6.0 or any of its non-Microsoft Corporation equivalents, such as Dr. DOS 6.0, and the Microsoft Windows 3.1 operating system. Most major computer manufacturers and dealers include these programs with a new system, and it's safe to suggest that in today's computer marketplace you should steer clear of any offer that does not include such a basic software package. Today's leading software makers are producing sophisticated programs with hefty price tags. A computer "buy" that excludes operating and other software usually bundled with a new computer is probably no bargain at all.

Apple Macintosh users will find their machines bundled with the proprietary System 7 operating system, which will read files from disks created on MS-DOS computers, and usually Microsoft Word 5.0 for MAC, the equivalent of Word for Windows, and Microsoft Excel 4.0, a very sophisticated spreadsheet program.

Word-processing software, such as Word for Windows, Word 5.0 for the Mac, WordStar 6.0 for

Windows, Ami Pro for Windows, and myriad others, allows the user to create professional-looking documents. Today's WYSIWYG (what you see is what you get) programs are a great help in creating professional page layouts for newsletters, books, manuals—you name it. If these programs are combined with desktop-publishing software, such as PageMaker or QuarkXpress, and a laser printer, documents can be produced that a few years ago were within the capabilities of professional typesetters only.

In the realm of mathematical software, including programs for accounting and creating spreadsheets and relational databases, there are even more choices, with Microsoft's Excel 4.0, Lotus 1-2-3, and literally hundreds of others competing for this segment of the business. There is a program available to address just about any user need. Pick up copies of the leading monthly computing magazines—for example, *PC Computing* and *PC Magazine*, for IBM, and *Macworld* for Mac users; they will keep you up to date on what the industry has to offer for your particular professional or business application.

A number of very sophisticated yet inexpensive and easy-to-use programs are available that combine some or all of the aforementioned functions. One of my favorites, and one I think is particularly well suited to onboard use, is Microsoft Works for Windows (a version is also available for MS-DOS). Works combines word-processing, spreadsheet, telecommunications, database, and file-manager programs in one integrated, easy-to-use, menu-driven program that enables the user to create, manipulate, store, print, send, and receive files and documents.

Computer-Assisted Communications at Sea

While at sea, the ability to access computer information databases and on-line services can add new dimensions of convenience, utility, safety, and recreation to the cruising lifestyle. To be able to send and receive documents; handle mail, messages, and financial transactions; update your float plan and ETA at the next port; order parts, provisions, or both; and get medical, financial, news, weather, and a plethora of other information while at sea would be considered a boon to all but the most traditional of bluewater cruisers. It is truly liberating not to be tied to the shore for anything, if you so choose.

It might be interesting to note that most of the initial research for this book was conducted on board a boat, by logging on to the CompuServe Information Service, a large on-line service based in Columbus, Ohio, and to Delphi, another on-line service based in Cambridge, Massachusetts. Among its hundreds of on-line "forums" and "infobases," CompuServe has a service called Sailing Forum. Delphi has a similar service, the Delphi Yacht Club, with members logging on from all over the world. Delphi subscribers also enjoy the advantage of being able to access, search, and retrieve data from the worldwide Internet, linking universities, governments, and commercial enterprises, about 15 million computer users in all. Use of this service is growing rapidly due to its very low hourly cost and the fact that, in addition to the marine special-interest areas, the service provides its subscribers with a full menu of news, entertainment, financial and business information, and access to a wide array of Special Interest Groups (SIGs).

Both U.S. coasts have large and active bulletin-board systems (BBSs) aimed directly at the boating market. As distinct from the commercial services, such as Delphi and CompuServe, these BBS services provide access to news and information about marine activities in specific regions of the country, and a message-exchange system between other members, both at sea and in ports throughout the world—in general, an information exchange among mariners who are knowledgeable about and interested in computers and boats.

In addition to being an excellent place to get boating information, on-line BBSs are also an invaluable source of public-domain data files and software programs that can be downloaded directly into your computer. Games, recipes, wine reviews, first-hand equipment and boat reviews, accounts of personal experiences in sailing areas around the world—a virtually unlimited list of topics—are available using on-line computer access and the growing shelves of "library" files on the services.

At sea, the cruising sailor or racer can gain access to these and a host of other telecommunications services via such utilities as C-Link by ComSat Marine Services, which provides satellite uplink services to onshore utilities. There is also the Trimble Navigation Galaxy Inmarsat-C system, a thorough description of which is found in Chapter 8. This system integrates the ship's GPS and Loran navigation units, weatherfax, and telecommunications systems. Autolink RT, an international service of Cimat Spa of Italy (Autolink RT stands for automatic direct dial maritime radiotelephone system), provides a radio hookup to the telephone networks of Great Britain and Europe.

Through these commercial utilities and new ones like them that are being introduced to the marine industry even as this is written, the onboard computer equipped with a modem can access other computers and networks through VHF, MF, and HF SSB radio transmissions. The Autolink RT models, which include handsets like small VHF radios for encoding the computer modem signals for transmissions between vessel and onshore stations, enable users to make direct-dial calls through international telephone networks, and to send and receive computer data via modem. The unit, which is basically a small, single-sideband radio with the added ability to encode data transmission signals for security reasons, is powered by a ship's electrical system (either 12-volt or 24-volt). After dialing and gaining access to a telephone network, the user can then log on to any of the world's thousands (with numbers growing) of on-line services and databases. Using Autolink RT and an onboard SSB radio set to access shore-based radio stations, a cruising navigator has direct-dial service from anywhere in the world to anywhere in the world.

There are some additional costs for this radio equipment. Although most yachts are already equipped with an SSB radio, Autolink RT radio units cost between $500 and $1,200, depending on the model and its capabilities. In addition, telecommunications charges as of this writing run an estimated average of $1.75 per minute, which can be billed to your telephone or credit-card account while underway. Considering, however, that an average call probably ranges from under 5 minutes to no more than 10, this cost seems small for the tremendous capability such equipment affords at sea. Furthermore, one might reasonably expect these costs to come down over time, just as did the costs of computers and computer components.

The (Not-So-Distant) Future

The boating community is just beginning to perceive the many ways computers can be used on board. The obvious applications, such as for navigation, pilotage, and record-keeping, were the first to emerge. Next will come the realization that the entire world is opened up to the boater through the small window of a computer screen. Details that once vexed the cruising life—mail and message handling, banking, shopping for boat equipment and provisions, getting news of the world while at sea, filing and updating a float plan, arranging for and maintaining vessel registry documentation, and obtaining information about foreign ports and entry and visa requirements—are being readied for easy handling while at sea.

There's even a program on the drawing boards for a computerized watch system that would enable the computer to stand a watch, taking some of the burden of this fatiguing and monotonous chore off the boat's crew. Imagine a vigilant electronic deckhand that will stand watch without bickering, falling asleep, falling overboard, becoming inebriated, steering off course, gybing, or waking others in the crew to ask for food and beverage! The trend is toward fully integrated systems that tie a computer with electronic charting and satellite-navigation programs into the autopilot. The information received by the computer via satellite will be transmitted directly to the helm, much as it is in a modern airliner. For a more extensive description of how these systems are integrated, see Chapter 8.

Just as computing is catching on with the sailing and power-boating communities, boating is definitely becoming popular with the electronic community of on-line services. Look for a rapid proliferation of computer-based information services and other resources as this trend reaches its full potential. Some of the existing services are already planning major expansions of their marine-oriented programs that will include more business services to the commercial marine industry.

Choosing the Seagoing Computer

The proliferation of computers, accessories, and peripheral components presents a dizzying array of choices for the boatowner who is considering an onboard computer installation. For anyone other than a diehard computer enthusiast or "techie," keeping up with changes in this technology and putting together a complete onboard computer system might appear to be a daunting task indeed. This book is dedicated, in large part, to sorting out the myriad choices and to helping the reader design an integrated system best suited to his or her needs and desires.

Ideally, a boatowner should strive to create a situation where the onboard computer becomes the ultimate tool, where there is a synergy among the crew, the boat, and the computer system. The seagoing computer must be accessible and configured so it makes a maximum contribution to life aboard.

The specific computer setup to achieve this ideal will vary, depending on whether the boat is a bluewater cruiser intended for long passagemaking, a coastal or day cruiser with less ambitious itineraries, a racer, a power yacht, or a boat that rarely leaves its marina slip. The degree to which the computer is expected to be used as part of the routine afloat will have a great bearing on the type of system the boatowner will select.

Computer Choices: Desktop Versus Laptop Systems

The first consideration in choosing a computer system for a boat is whether to select a full-size IBM PC (or compatible clone) or Macintosh desktop, such as one you might already use in the office or at home, or a much smaller, less space-hungry laptop or notebook computer. There are many pros and cons to evaluate in making this important decision.

Size, obviously, is one of the major differences between these two options, but it is not the only one. The two types of computers have different capabilities, strengths, weaknesses, and maintenance and durability considerations. They even have different levels of subjective values, such as what is often referred to as user-friendliness, or physical features, such as the feel of the keyboard and the readability of the monitor screen.

At present, computer manufacturers seem to be concentrating on addressing those areas in which the notebook computers are overshadowed by larger systems, namely screen-display readability, battery-charge duration, disk-storage capacity, and keyboard size and feel. Whether the notebook computer will ever be considered sufficient for all-around use and as the primary computer for all needs is a subject that has been debated endlessly—and no doubt will continue to be for some years to come. But it is fair to say notebook-computer technology has been the area in which the greatest advances have been made in the past year. The whole setup—from screen, keyboard, and disk drive to overall size and weight—has undergone a radical transformation. The weight of the smallest notebook computer has dropped to as little as 1.3 pounds for such units as the $595 Zeos Pocket PC. In another development, Apple Computer has introduced the Apple "Newton," a Star-Trek-esque shirt-pocket computer and telecommunicator capable of reading input from handwritten notes on a screen using a stylus-type pen. It is equipped with word processing, spreadsheet, and telecommunications/fax modem programs and costs just under $1,000.

Computers specifically designed for onboard marine use have only very recently appeared on the scene. Up until now, most installations have been adaptations of systems designed for home or office use. Marine users may now choose from among computer systems that are somewhat hybrid in nature, different from standard desktops in their housing and their configuration but distinct from the popular laptops in terms of addressing common complaints about screen displays and adaptability to the demands of the marine electronics station.

The day is fast approaching when a notebook computer with a state-of-the-art 486 or 68030 processor chip, 4 MB or more of RAM, a 120-MB hard disk, a VGA or color VGA screen, and a full-size keyboard may be all the computer most people will need. I'm inclined to think this standard will apply to marine computer users to an even greater degree. Despite the minor compromises involved in using the notebook computer on board, the advantages are indeed considerable.

The Full-size Desktop System

The conventional, full-size IBM PC or compatible and the Apple Macintosh consist of a CPU, a CRT monitor, and a typewriter-style keyboard (Figure 2-1).

Figure 2-1. The conventional PC may be installed aboard a boat in its unaltered, out-of-the-shipping-carton form with few modifications.

Altogether, these components take up several square feet of space, but generally they are manufactured and sold as a system consisting of three separate components that can be mounted separately to maximize use of precious cabin space.

The majority of onboard computers are mounted in the navigation station or chart-table area. When a desktop computer is installed, the CPU is generally hung in a rack or mount beneath the table or built into a mounting somewhere in the nav-station cabinetwork. Besides maximizing use of space, this type of installation provides for a secure mounting yet affords the necessary accessibility to the computer-system components. It is necessary for the CPU to be mounted so the front, which houses the floppy-disk-drive slot(s), is within convenient reach. Controls such as the on/off switch, keyboard lock, and reset button are also sometimes mounted in the front of the CPU. The monitor can rest on the nav table or be built into the cabinetwork; the keyboard can rest on the chart table or be moved about at will.

For long work sessions or for viewing on-screen navigation charts, weatherfaxes, tables, lists, and particularly any kind of computer-aided-design (CAD) work, such as architectural or engineering applications, the full-size screen of a desktop system can be a major advantage over the smaller screens of the popular laptops. Today's full-color, high-resolution graphics-display adaptors, with which the vast majority of desktop computers are sold, produce sharp, stunning images at a relatively low cost. To obtain a display with equivalent quality in a laptop or portable computer involves a considerably higher cost. Generally, miniaturization involves costs that are inversely proportional to size. A two-to-one ratio may be employed as a general rule when estimating the relative cost of a desktop IBM, IBM clone, or Macintosh computer against the cost of a similarly equipped and capable laptop. This is expected to change as technology develops and the cost of miniaturization continues to fall.

The full-size CPU of a desktop system has some advantages as well. Although it takes up much more space, its case can be opened more easily, and the parts inside are easier to get to and work on when necessary. In the interest of seagoing self-sufficiency, most add-on components, such as modems, additional memory, and disk drives, can be installed or repaired by the user while underway. In contrast, the size constraints of the laptop computer have resulted in extreme miniaturization of components and dense packing of chips, processor boards, mini-modems, internal components, and accessories. Delving into the internal workings of these computers practically requires the steady, precise hand of a jeweler—a task not recommended in a bounding sea at 20 degrees of heel.

There is little argument that the keyboards of most desktops have been easier to use than those on a laptop. There are exceptions now, including the Zeos, Dell, and Toshiba lines of laptops that boast full-size IBM-Selectric-style keyboards with either 88 or 101 characters (Figure 2-2). In general, the full-size, detached, 101-key desktop keyboard is favored by most heavy computer users. Again, there are space considerations involved, with a desktop keyboard and monitor taking up a great deal more real estate on a nav table than an integrated keyboard and display on a small laptop computer.

Finally there is the question of cost. The prices of both full-size and laptop computers have declined dra-

Figure 2-2. Laptop computers such as this Toshiba 3300SL not only have the bright VGA backlit display sought by most of today's computer users but the full-sized keyboard too.
(Photo courtesy of Toshiba America)

matically in the past five years, and they continue to fall. This price reduction has come about primarily as a result of the proliferation of IBM-compatible clones in the marketplace, competition in patented technologies and the resultant fall in the cost of the silicon computer chips that serve as engines for all computers, and the huge increase in the volume of computers built and sold. In the case of the Apple line, competitive forces have driven prices of Macintoshes, including Apple's great series of Power-Book laptops, down closer to the cost of IBM compatibles.

But, as mentioned earlier, due to the still much-higher cost of miniaturization technology, laptop computer prices have remained proportionally higher than those of conventional, full-size computers. The onboard computer system, exclusive of highly specialized peripherals such as a radio modem and satellite telecommunications equipment, can be purchased and installed for under $3,000, sometimes for considerably less. However, in considering the price differential between a desktop system and a laptop, the prospective buyer can figure on the difference between comparably equipped systems at about a two-to-one ratio, the buyer paying more for the more sophisticated technology embodied in the laptop.

The New Laptop and Notebook Computers

The convenience of a small computer that will fit in a briefcase or drawer, weigh less than six pounds, run on either an internal battery or ship's AC power, and have most, if not all, of the display and disk-storage capabilities of a full-size desktop system is almost irresistibly appealing. Advances made in today's tiny laptops have brought them closer to the capabilities of a full-size system. Tiny, large-capacity hard disks; elegant, backlit VGA screen displays; and large-capacity, nickel-cadmium (NiCad) or nickel-hydride (NiHi) batteries have become the rule rather than the exception in laptops. Today, approximately 60 percent of the computers installed aboard yachts are laptops. For the moment, laptops represent a fine compromise.

A Comparison of Two Marine Computer Systems

The two systems presented in detail below will give you an idea of what's available in the computer-hardware systems designed for the marine market. It is by no means an endorsement of any particular product or design (nor is any mention of a commercial product in this book). While the manufacturers mentioned here have been extraordinarily helpful in providing product and technical information, pricing, photos, and diagrams of their wares, their products are included here as representative of a trend. There will be more computer products available to the marine market, some better and some worse. In the end it boils down to the age-old warning: *Caveat emptor.* You must analyze your needs, your budget, and the products available, and then decide for yourself what is best for your purposes.

Let's take a look at two computer systems designed especially for the marine market: the Sea PC marine computer, a full-size desktop, and the Ultrathin 386 SL notebook. Both are designed, manufactured, and sold by D.F. Crane Associates.

Sea PC

The Sea PC is designed and manufactured with the rigors and demands of the marine environment in mind. Incorporating features that address the requirements of seagoing computers, this desktop component system is small enough to be installed aboard almost any size vessel and can be taken home or to the office in an optional soft carrying case (Figure 2-3).

According to Dave Crane, designer of the Sea PC, its components are waterproofed using a proprietary process to seal the internal boards and other electronic workings of the unit. Tightly sealed cases enclose the CPU, monitor, and keyboard. "The Sea PC is 'splash-proof' and has been 'ruggedized' to withstand the harsh

Figure 2-3. The SEA PC's 18-gauge-steel enclosure and Plexiglas front door ensure a moisture-proof environment for the computer's critical internal components. (Photo courtesy of D.F. Crane Associates, Inc.)

demands of marine service," Crane says. "We've responded to the demands of the marine market with a durable PC, incorporating advanced technology in a stylish design, and offered at an affordable price. All its features—design, size, price—combine to make it an ideal choice for those looking for the 'right' computer for their boat."

The Sea PC features separate, smaller-footprint components that are designed to be flexible both in how they are mounted and in their power requirements. The Sea PC can be mounted horizontally or vertically in the nav station, inside a locker, or attached to the overhead. A small DC-AC inverter is built into the console, allowing the system to accept either 12-volt DC or 110-220 VAC

Figure 2-4. The smoked-Plexiglas door of the Sea PC enclosure allows the user, but not the elements, access to the on/off, computer reset, and turbo switches, as well as to the floppy-disk-drive door. (Photo courtesy of D.F. Crane Associates, Inc.)

power. The Sea PC uses only 24 watts for the CPU and 54 watts for the Super VGA color monitor. This translates to a power draw of less than 1 amp on AC or 66.5 amps on DC for the total system.

The CPU The enclosure for the CPU is an 18-gauge-steel-plated protective case that is completely sealed against the environment. It is zinc-dipped, filed, primed, and coated with a baked-on finish that resists salt air and moisture (Figures 2-3 and 2-4). The unit measures 6½ inches high by 4½ inches wide and 16½ inches long and weighs 14 pounds. A splashproof door allows easy access to the basic controls and 3½-inch, high-density floppy-disk drive. The entire CPU and 120-MB hard disk (expandable to 240 MB) are isolated from vibration and shock.

At the heart of the machine is a full-size, high-performance 386SX computer processor that has 2 MB of RAM (expandable to 5 MB) and operates at a reasonably fast 25-MHz clock speed. This is more than adequate to enable the computer to run multiple programs utilizing file- and task-swapping in a windowed graphics environment.

Operating Software The Sea PC comes preinstalled with the latest version of the MS-DOS operating system (MS-DOS 6.0), QEMM memory management, and DESQview/386 software. DESQview/386 is as easy to use as Microsoft Windows. It also allows almost any program to run in a true multi-tasking windows environment, unlike Microsoft Windows, which requires software to be designed to run under its own operating system. Also included is "Checkit," diagnostic software that helps identify the causes of any hardware problems.

Color Monitor The Sea PC's Super VGA color monitor measures 10 inches wide by 9 inches high and 12 inches deep and weighs 16 pounds. It will fit into even the most cramped navigation station. Its graphics display has a resolution of 1024 x 768 pixels and a .28-inch dot pitch, resulting in brilliant, sharp color definition. This is particularly important for graphics programs like weatherfax and electronic navigation charts. The monitor can be installed up to 32 inches away from the CPU.

Serial and Parallel Ports The Sea PC has four serial ports. This is one requirement in particular in which a seagoing computer must differ in design from its land-bound forebears. Nautical applications such as weatherfax, charting, navigation, GPS/Inmarsat, and telecommunications require more serial I/O ports than the one or two found on conventional personal computers. One port is allocated to the computer's trackball. The others can be used for a weatherfax, for input from a navigation receiver, for a serial printer, or for components to be added sometime in the future. One parallel port, for use by a parallel printer, is also provided.

Keyboard and Trackball   The compact, 84-key keyboard measures 6½ inches by 13 inches and includes a Microsoft Ballpoint mouse pointing device. The pointing device, also known as a trackball, can be attached to either the right or left side for use by either hand. A clear, flexible, vinyl "skin" covers the keys perfectly to seal out dirt and moisture.

Warranty and Repair   The Sea PC is typical of most computer systems today in that it comes with a full, one-year limited warranty. If a hardware problem crops up, the proprietary Sea PC diagnostic software checks out the system and helps determine whether the problem is in the CPU, monitor, or keyboard. The faulty component is then shipped to the factory in San Diego and D.F. Crane repairs or replaces it and ships it back, usually using next-day air service. For systems under warranty, the customer pays only for shipping; for systems out of warranty, the customer pays for parts and a nominal $49 repair charge, plus shipping. According to D.F. Crane, repair service is available anywhere in the world.

The Sea PC sells for $2,995.

Ultrathin Notebook PC

The Ultrathin notebook PC is an ordinary-looking notebook computer designed for seagoing use (Figure 2-5). It employs an external hard disk that is easy to remove and take ashore. The entire unit is 1½ inches thick, 8½ inches long, and 11 inches wide, and takes up less space than a ream of letter-size paper. The unit goes light on gross tonnage too, weighing a mere 4½ pounds.

The Ultrathin's CPU contains a 386SX processor running at 25 MHz and comes with 2 MB of RAM

Figure 2-5. The Ultrathin Notebook PC, aimed directly at the onboard-computer user, is 8½ inches wide and 11 inches long—the size of a standard piece of typewriter paper—and weighs in at 4½ pounds. (Photo courtesy of D.F. Crane Associates, Inc.)

(upgradable to 4 MB). The computer's display is a paper-white backlit LCD that displays 32 gray scales with a resolution of 640 x 480 pixels. The Ultrathin can display sophisticated graphics programs, such as nautical charts and weatherfax, with the ease and accuracy of a full-size desktop PC or Mac.

The price of the Ultrathin 386 SL at this writing is $1,295. A new 486 SLC model is now available for $1,795.

Other Considerations

Processor Type and Speed

Important considerations in choosing an IBM or compatible computer are processor type and speed. There is a tremendous range of choices and capabilities, and the marketplace is awash with esoteric jargon pertaining to them. The technical hodgepodge basically boils down to this: The state-of-the-art IBM or compatible computer runs on either an Intel or equivalent 80386 or 80486 processor chip, or their "SX" offspring, with a "clock speed," or measurement of the computer's data-processing rate, of at least 25 MHz.

A word is in order about the distinction between the "SX" and "DX" designations after some computer-processor labels. In simple terms, an SX chip is a somewhat toned-down version of the "full," untruncated DX chip, having slightly less processing speed and capability. Average users might never notice the difference between a computer running on an SX and one with a DX chip, unless they are using very large or complex programs.

There are thousands of computers on the market that contain chips other than an 80386 or 80486, such as the older Intel 80286 or the even older 8088 or 80C8. To put it bluntly, these processors are obsolete. In fact, by the time you read this, even computers based on the Intel 386 chip might be considered obsolete, replaced by computers with the 486 or even the 586 chip. The 586 will be known as the P5 or Pentium, according to Intel Corporation.

Do not succumb to the temptation to buy an 8088 or a 286 machine, regardless of advertised "unbelievably low discount prices." Today's sophisticated software, including programs written for marine applications, simply will not run on these older, smaller, slower computer platforms. Nowadays, as you'll see in the comparisons later in this book, you really aren't paying much more for faster speed, and the difference between a 12- or 16-MHz 286 system and a 20- to 33-MHz 386 or 486 system is like the difference between a Yugo and a Ferrari on the autobahn. A bad decision here will haunt you again and again as you try unsuccessfully to use various software programs on your machine.

For those who choose an Apple Macintosh system, the processor-speed consideration is less important—at least for the moment. Macs run on the Motorola 68000 processor chip set, which boasts both speed and brawn and is one of the reasons why the Mac's price is substantially higher than that of an IBM clone. Current desktop Macs use the 16-MHz 68030 chip, for the entry-level Classic, to the 33-MHz 68040, for the Quadra 950. The Mac PowerBooks use 25-MHz to 33-MHz versions of the 68030.

System Memory and Disk Storage

All of today's computers are hungry for two things: memory and storage. Here again we find an area where technology has bounded ahead at a dizzying speed, creating new standards and rendering some systems obsolete.

System memory, commonly referred to as RAM, can be thought of as how large a "workroom" a computer has available in which to operate and to run programs. Simply put, RAM is installed right on the "motherboard," or main board, of a computer in the form of silicon chips.

Disk storage is the amount of space available to the user for storage of files and programs on devices such as internal hard disks and floppy disks.

In the early 1980s, 64 kilobytes (K) of RAM and disk storage was about all you could get. There were no hard disks then, only "floppies." The first hard disks were introduced in the early- to mid-1980s and had a capacity of about 8 to 10 MB, which was considered more than enough since it could store an entire recipe file, a master's thesis, and the draft of a Tolstoy-size novel. Today, however, when a sophisticated operating system and word-processing program together can occupy nearly 20 MB of disk space, it's hard to imagine having too much. Eighty MB is about the smallest hard disk one should consider, and hard disks of 660 MB or even a gigabyte or more are not uncommon—but neither are they necessary for most applications.

Today, with Mac or IBM, it is difficult to operate with less than 4 MB of RAM, and it is desirable to have 8 MB or more. Although it is easy and relatively inexpensive to add RAM by plugging modular SIMM or DRAMM chips into expansion slots in the computer's main board, it is by far preferable to buy a computer with 4 MB (or more) of RAM and be done with it. Most new systems are sold with 4 to 8 MB of RAM. You may add storage capacity—up to a total of 32 MB—on most Intel 386 and 486 boards. Unless you are running complex computer-aided-design (CAD) programs, or extremely complex graphics or spreadsheet programs, it's doubtful that you'll require more than 4 to 8 MB.

Having an adequate amount of RAM installed on your machine will also add to battery life and endurance, since the computer's CPU will spend less time accessing the hard-disk drive while managing software programs. Hard-disk drive access is a large drain on the battery.

Disk storage is more difficult to add to a computer later on, so it is desirable to start out with a computer with a large-capacity hard disk. Large-capacity hard disks are available on both standard-size desktop computers and laptops.

To put this discussion of storage into perspective, the popular word-processing program Microsoft Word for Windows (version 2.0), along with its operating system, Microsoft Windows (version 3.1), occupies 13 MB of hard-disk storage space. When operating, the program uses up to 2 MB of RAM as it goes about its various tasks of displaying menus, swapping programs and files back and forth, and manipulating data. The text of this book amounts to a little over 2 MB of data.

One final comment about these considerations of RAM and hard-disk storage: There are no physical size or dimensional differences in different RAM or disk-drive capacities. A 210-MB hard disk is the same physical size and shape as a 40-MB hard disk, and RAM uses internal CPU board space already allocated for the purpose. "Bigger is better" where RAM and hard-disk storage capacity are concerned.

Installing the Seagoing Computer

Planning the Installation of an Off-the-Shelf Desktop Computer

We've already noted that the installation of a computer system aboard a boat can be as simple as carrying a notebook computer on board and placing it at the navigation station. For our purposes here, however, we will assume that the boatowner has decided to install a state-of-the-art IBM PC or compatible 386 or 486 computer, along with GPS, autopilot, weatherfax, and full word-processing, spreadsheet, fax, and printing capabilities.

The ideal installation takes into account all factors involved: available space, integration with other electronic components, the boat's electrical power, ease of use, and access to the machine for repair and maintenance.

On this latter note, it is worthwhile to mention that the incidence of problems in modern computer equipment has been significantly reduced over the past few years. With the advent of myriad advances, including smaller, faster, and more rugged disk drives, internal components that produce less heat and are less sensitive to dust and motion, and sealed circuit boards, computer installations can go years without a single glitch. Today's computer problems are more likely caused by software bugs and the integration of peripherals than by the functioning of the computer itself. Care must be taken, therefore, to put together a system that works smoothly with other boat equipment so that the owner and crew are not burdened with hardware-compatibility problems.

If you are "spec"-ing out a new boat, make sure the designer or builder makes adequate provision for the computer installation. In fact, many yards will even install a computer system for you. For the rest of us, as with so many other aspects of the nautical life, careful planning and improvisation are called for. Before a trip to the local computer store or a call to one of the many mail-order computer sellers, it is important to look over your boat's layout below and carefully think through any possible size and space restrictions. For example, if your chart table is a fold-up type with only a couple of square feet or so of surface area, you will not want to install a computer with a monitor that not only takes up the entire chart table but can't be mounted anyplace else. If every available inch of locker and cabinet space is already used, decide if you want to make room for a CPU mounting or whether you need to consider a notebook computer (or another boat!). Take some measurements and compare the space you have available for a computer installation with the specifications of the particular computer(s) you have identified as meeting your needs. Most, if not all, computer manufacturers include the dimensions of their equipment in their advertising. Thus you can quickly determine if a particular model won't work for you.

Once you have sized up all the physical requirements of the computer installation and bought a system from a store or through the rapidly growing array of direct-mail computer sellers, such as Gateway, Dell, Zeos, CompuAdd, Zenith Data Systems, and hundreds of others, you are almost ready to install your system.

First, however, you must evaluate the structural elements of your installation. Basically, this evaluation amounts to ensuring that the structural parts of the boat to which you intend to mount your computer components are capable of sustaining the loads. You can hire a marine architect or engineer to do this for you, something few boaters of my acquaintance would be willing to pay to do, or you can do it yourself, letting common sense be your guide. Find suitable bulkheads, joinerwork, or cabinetwork that will bear up comfortably under the load of a 15- to 20-pound CPU in rough weather. Don't attach CPUs or even the vertically constructed "tower"-type cases to chart-table legs, for example. Table legs can't be relied upon to handle the acceleration loads imposed on them by the combination of computer components and a pitching boat.

Nor will you want to leave computer components unattached or merely wedged into the boat's cabinetwork, unless you plan to stand hard-by the marina slip.

Configure for the worst-case scenario: pitching, rolling, heeling, bounding—even knock-downs—and make sure every component of your computer system can be fully secured. I've heard many a seagoing-computer horror story about flying components and machines smashed to worthlessness on the high seas. A bit of extra care and planning here will pay off handsomely at sea.

Fastening Computer Components

One of the best ways to fasten a CPU to the boat is to use ¾- or ⅞-inch stainless- or galvanized-steel straps, or stainless-steel mounting brackets. These fastenings are impervious to rust and corrosion and designed to permit a tight fastening (Figure 3-1). They are also relatively inexpensive. Fasten the CPU in at least two places, securing the straps or brackets to the boat with AN-quality anodized bolts at least ⅜-inch long. Use either neoprene Sta-Lock–type nuts or, better yet, neoprene Sta-Lock–

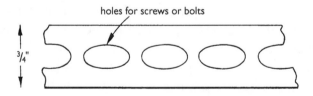

Figure 3-1. Flexible galvanized straps are inexpensive, strong, and easy to use to fasten a CPU to the boat.

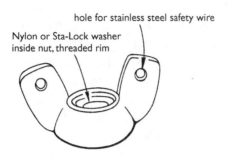

Figure 3-2. An anodized wing nut with a Sta-Lock washer makes for easy attachment and removal of your CPU.

type wing nuts for easier attachment and removal (Figure 3-2).

Nylon straps can be used in lieu of metal straps for fastening a CPU to a boat. However, nylon straps tend to come loose underway and are therefore not as reliable as metal fasteners. They do have the advantage of being lightweight, versatile, and easy on joinerwork. I've seen all manner of fasteners used to hold computer components, including C-clamps and even bungee cords—these are definitely not recommended, for obvious reasons. A small C-clamp can be used to secure a monitor to a chart table, if necessary.

The best solution for mounting a computer keyboard, aside from buying a laptop where the keyboard is an integral part of the machine, is to mount it on a drawer-type tray that slides under the chart table. Not only does this type of mounting allow for the safe, convenient stowage of the keyboard when not in use, but it places the keyboard at a level below the table top that is much more conducive to comfortable typing.

Before drilling the first hole, however, check to make sure that all components can be reached by electrical cords, and that all cables for connecting computer components, such as keyboard, printer(s), monitor, fax machine, radio(s), and modem, can reach the CPU. Also, bear in mind that you might have to disconnect and connect these items frequently to modify or maintain your installation. Consider the computer's proximity to the boat's power panels, electrical distribution buses, and outlets. In most cases, the navigation station is the area closest to these utilities.

Electrical System Integration

If the computer system's power cords don't reach the boat's electrical panel or outlets, the problem can be solved easily using a multiple-plug surge protector with an extension cord. Surge protection is advised for all computers and peripherals anyway, and using one accomplishes a couple of things at once. Most bus-type multi-plug surge-protection units allow flexibility in the location of equipment up to ten feet or more from an electrical outlet (Figure 3-3). Surge protectors also shield fault-sensitive electronic equipment from fluctuations in the phase and voltage from a power source. Since marine power—whether 110–120 VAC shore power or onboard 24-volt DC inverted to 110–120 VAC power—is more prone to surges, faults, outages, and cycling on and off than stable household or office electricity, a surge protector is an absolute must. Don't even think of operating without one.

Installing the Seagoing Computer   45

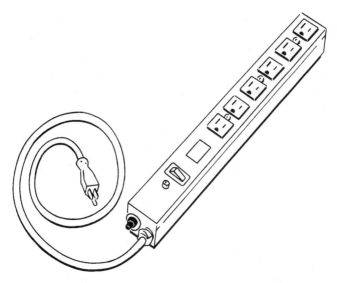

Figure 3-3. This typical surge-protection unit enables the user to plug in several appliances at once and also increases the distance from the main electrical power source that the computer and its peripherals can be mounted.

Be sure, however, to get a surge protector capable of handling plenty of amps, or you might defeat your purpose. Add up the amperage draw of each component to make sure that the total does not exceed the maximum rated amperage of the surge protector. These values are almost always clearly listed on the back of the electronic component. As an example, a typical surge protector might be rated at 10 amps. A typical IBM-compatible 386DX computer draws 5 amps. Its VGA color monitor draws 1 amp. Right there you have 6 amps, with 4 amps to be divided among all other peripherals. A stereo can sop up 2 amps, a color TV draws 4 amps, and a standard Loran, an SSB radio, and a VHF transceiver use 1 each—so exercise caution.

The surge protector itself can be stowed out of the way once the various power cords are attached. There should be little need to access the unit regularly. However, examine your surge protector to determine if conventional plug-in-type fuses are incorporated into its design. If so, there may be an external fuse that will need replacing if it blows. Newer designs, as a general rule, do not use external fuses. Some designs employ indicator lights that illuminate when the device has been activated by a surge or power failure; some require manual resetting after activation, like a circuit breaker, to prevent repowering the unit and the appliances attached to it before the cause of the fault has been corrected.

If your boat is equipped with a DC-AC inverter, you might already have surge protection—up to a certain rated limit. The protection incorporated in the inverter may eliminate the need for additional surge protection, but be sure to check the maximum amperage load the inverter is capable of carrying.

Power Requirements

Computers and their peripheral accessories operate on alternating current (AC) power. This is the type of power found at the ubiquitous electrical outlet in homes and offices. It is also available as shore power at marina slips for any boat equipped with a shore-power AC system and the necessary AC-distribution panel and outlets. Most boats are so equipped today. For a vessel equipped with shore-power accessories, the installation of a computer and peripherals presents no problem until the boat is disconnected from shore power in preparation for getting underway. The ship's primary generator and storage batteries provide direct-current (DC) power. To have a supply of AC power for running appliances, a boat must be equipped with an inverter, a portable generator with AC-output capability, or a permanently mounted gen-set.

DC-to-AC Inverters

DC-to-AC inverters change 12- or 24-volt DC current into 110–120-volt AC electricity for use by household appliances, computers, and peripherals. These units vary from small-output devices with a maximum rated output of under 200 watts to large systems of 2,500 watts or more.

The computer system with a CPU, CRT, and modem uses 250 watts of 110 VAC power, or about as much as a large reading lamp. Printers and other peripherals require slightly more, but the typical system probably uses no more than a total of 300 watts.

A small inverter, such as the PowerStar (Figure 3-4), is a good solution for the problem of converting power from DC to AC for a seagoing computer installation. The system can be wired directly into the ship's DC power system or hooked into it through a cigarette-lighter-type plug. The PowerStar produces up to 400 watts of AC power, with 140 watts of continuous output, and also functions as a surge protector that isolates a computer and its related accessories from irregularities produced by a boat's DC power system. Such small units, weighing in at around one pound, represent the low end of the inverter spectrum at a cost of about $150.

Figure 3-4. The PowerStar 200-watt inverter.

A boatowner can select from a wide array of systems that not only provide more power output for additional appliances but can be equipped to serve as battery chargers as well.

Heart Interface manufactures large inverters with outputs of 600, 1,800, and up to 2,800 watts. Prices begin at around $1,500. Other large AC inverters are available from Trace Engineering, Professional Mariner, and Statpower.

A quick survey of all electrical equipment on board that might be powered at one time by a marine electrical alternator or inverter will indicate how big a system your boat requires.

Safeguarding the Computer Installation

As with all marine systems, special precautions must be taken to safeguard a seagoing computer from moisture and other potential sources of damage. Direct contact with water, dust, salt, and pollution-laden air, excessive motion, exposure to projectiles and other collision hazards, magnetism, heat and cold, and other onboard hazards are of special concern. It is wise to recognize and take as many of these potential hazards as possible into account in the process of installing the computer.

Some of the obvious pitfalls can be the most difficult to avoid. For example, a great many navigation stations are located near the cockpit or main cabin hatch. This is great for dashing below to look at a chart or weatherfax, but it's also the one place below most likely to get soaked in heavy seas or a sudden cloudburst. Try to locate the computer and its components where they will be least likely to get drenched in the worst-weather scenario.

By the same token, locating computers directly below ports and other hatches is not a wise move. Computer monitors have been destroyed by cabinet doors opening into them in rough weather. Similar losses have resulted from flying projectiles in rough seas. If you cannot mount your system where it will be safe from all these potential hazards—a difficult, if not impossible, objective—at least have a system for safely stowing and battening down the most vulnerable components, including CRTs, notebook-computer units, and printers, during rough weather in a seaway. Custom-built covers or the ability to stow the machine in a locker during heavy seas are two of the better solutions.

Various covers are available commercially for computer consoles, CRTs, CPUs, keyboards, and printers. The new marine computers are already enclosed in waterproof, extra-heavy-duty casings. The merits of the Sea PC have already been mentioned.

Another computer system impervious to the harsh elements of seagoing life is the Bison Explorer Portable, a powerful, 32-bit, IBM-compatible 386/486 machine with a completely enclosed and integrated CPU, monitor, and keyboard (Figure 1-5). I don't know if this machine is bulletproof, but it's at least moisture-proof. Its designers and builders created the 24-pound Bison Explorer to withstand the rigors of gathering scientific Arctic data. It is waterproof and designed to function at temperatures ranging from +45 to −25 degrees C. It may not be an attractive-looking alternative, but it is a rugged one.

Zeos International, a major manufacturer of mail-order PCs, is another company that is addressing the special problems of marine computer applications, and the U.S. Navy is one its largest customers as a result. All Zeos conventional PCs contain specially weatherproofed internal components, and the CPU cases are sealed using a unique tongue-and-groove system that makes them more resistant to outside elements. This is done not only to protect the units from moisture but also to meet the company's rigid standards for reducing radiation emissions from all of its computer workstations.

Despite the logical incompatibility of electrical components, high-precision, high-RPM disk drives, magnetic media (floppy disks), and the wonderful mix of salt air and ubiquitous air pollution, today's computers are amazingly tough. Several years ago, there was an inordinate amount of concern about dust and moisture around computers. There was even talk of special dust-free rooms in office buildings where computers were used. Today's computers are capable of functioning with minimum difficulty in most environments tolerated by humans. The computer system on which this book was written is a good example of a trouble-free setup. It is a IBM 386DX clone, and it was set up in a home office with no special measures taken to protect it. I turned it on one afternoon in May 1990 and it has run continually since then, with the exception of a few unexpected power outages. Even those did not damage it, since the unit is adequately voltage- and surge-protected. The unit is being replaced, but not without a lot of admiration for its unstinting service. I hope the new one will do as well.

I have also had a fully integrated system on board my boat for two years now with no environmentally induced problems. This is not to minimize in any way the adversities under which the seagoing computer must operate, but only to help assuage any latent concerns that a marine environment is in any way too harsh for today's computer equipment.

I also mention this rather admirable durability of computers in order to dissuade boaters from taking extraordinary measures to try to weather- or waterproof computer components, especially internal components. Stories abound about boatowners sealing computer CPUs in plastic bags or cellophane, lining the insides of CPUs, printers, and other equipment with aluminum foil, or using special waterproofing aerosol sprays inside cases or on computer boards and accessories. These measures and others like them are apt to do more harm than good and could void your manufacturers' warranties. Remember that the insides of these boxes are built to very exacting standards, and that computer equipment generates heat and must have adequate air circulation to function. Also, do-it-yourself measures to seal out moisture can, in fact, have the opposite effect, sealing in moisture to short, corrode, and/or destroy the interior workings of your expensive new machines.

With proper installation, normal precautions while using the machines, and care in supplying electricity to the system, a mariner can expect to get years of trustworthy service from the well-designed and well-built modern computer.

Lighting the Computer Console Area

Lighting is an important consideration in any work area, and a factor too often neglected in setting up computer workstations.

A monitor's internal lighting system is designed to be used under optimum ambient light conditions. While all computer monitors have controls for adjusting the brightness and contrast of the screen display, these alone cannot overcome the glare from a screen bathed in direct sunlight, or offset the effects of a harsh fluorescent lamp. Nor is it a good idea to work in a dark room looking into the bright computer screen. Doing so for extended periods of time will cause eyestrain.

The best lighting system is one that you can adjust for changes in natural light when working at the computer. If the light's direction and intensity can be changed, the same system can be used for general lighting purposes.

If you are installing a computer under built-in lamps, such as those often recessed into salon or nav-station cabinetwork, try to adjust the console so the light from the lamps does not shine directly on the computer screen. Fluorescent lights create a particularly unpleasant glare on computer screens. Some built-in lamps are equipped with rheostats or on/off switches that will enable you to dim the lights or turn them off when using the computer screen.

Gooseneck or other flexible lamps are good to use with computer setups because the light can be moved to optimize lighting conditions. I use a flexible-arm light with a relatively low-powered 25- to 40-watt bulb at my workstation. This combination throws just enough light onto the computer screen and keyboard to balance the internal lighting of the computer monitor.

Low-cost brass nautical hurricane lamps can also produce a fine subdued light for a computer work area. They are a tasteful addition to any yacht "office" and have the added benefit of offsetting the high-tech appearance of the computer system with more traditional accessories.

4

Choosing and Linking Computer Peripherals

Modem and Fax

Perhaps the greatest enhancement to the utility of a seagoing computer was the invention of the modem, especially now that computer data can be transmitted via satellite and cellular telecommunications systems. The ability to transfer data from one computer to another is almost as important as the ability to use the computer for word processing and complex mathematical tasks.

Huge strides have been made in this area in the past five years. The first modems were big external boxes capable of transmitting machine data only at a snail-paced 300 baud (bits per second). Today, tiny internal modems fire files from a desktop PC, Mac, or notebook computer across the phone lines and back and forth through satellite uplinks at speeds of 14,400 bps (bits per second) and up. In addition, new hardware and software products have been released to keep pace with the growing popularity of fax.

Seagoing computer users can choose many accessories and enhancements for their onboard desktop or notebook computers, but a modem is virtually a necessity. As a practical matter, very few notebook computers are sold without an internal modem. High-speed internal and external modems are available as add-ons to all modern IBM-compatible and Macintosh computers. Prices generally range from about $129 for 2,400-baud internal modems without fax send/receive capabilities to as much as $600 for 14,400-baud V.32bis internal and external fax modems. Many of today's more sophisticated modems incorporate both high-speed data transfer and fax capabilities. The fax has become the method of choice for transmitting everything from a lunch order to a nearby deli to a 40-page proposal complete with spreadsheets and graphics, and it seems to be getting more and more difficult to live without one.

Typical of the modems available today is the Supra Fax Modem V.32bis (Figure 4-1), retailing for about $400, with an internal version for slightly less. The unit is fully "Hayes compatible," which simply means that it functions like the Hayes-brand modem that has become an industry standard. The Supra can send data at speeds of from 300 to 14,400 bps; with its new data-compression technology, it achieves an effective throughput of up to 59,700 baud. It is capable of sending a fax from your computer to any phase-three fax machine (the industry-standard fax-receiver configuration). With modem soft-

Figure 4-1. A new-technology lightning-fast Hayes-compatible external modem, the Supra Fax Modem 14.4/V.32bis. (Photo courtesy of Supra Corporation)

ware like Winfax Pro Version 3.0 by Delrina Technologies, the Supra (or any Hayes-compatible modem) can receive faxes just as an ordinary fax machine can, and store them on disk for reading (and editing, if desired), or for printing out on a laser printer.

With this capability, and with optical character recognition (OCR), which allows a computer to read all manner of text, graphics, and difficult-to-decipher characters such as handwriting, signatures, etc., a computer and laser printer can quite easily suffice in lieu of a stand-alone fax machine.

Given the choices available to a boatowner contemplating the purchase of a modem, an internal modem is the most logical choice, considering the need for space conservation on board. An internal modem mounts in an expansion slot inside the CPU case, making it out of the way, less vulnerable to damage, and not difficult to remove if repairs are needed, and eliminating the tangle of phone wires and power cords. And internal modems tend to be a few dollars cheaper because they don't need an external case.

The Cellular Modem Connection

A standard telephone modem, connected to telephone lines via the familiar RJ-11 phone jack, is fine for use when a boat is in port. Most marinas offer telephone connections to boaters at their slips. But what about while underway? That problem has been addressed through the use of cellular modems and satellite communications. Some notebook computers, such as Toshiba's T1800 and T3300, offer optional cellular modems that will transmit and receive both data and

Figure 4-2. The Toshiba 1800 Satellite is a state-of-the-art notebook computer capable of using the T24D/X cellular modem. (Photo courtesy of Toshiba America)

fax (Figure 4-2). This new technology sounds expensive, but relative to the cost of similar computers only a few years ago, it really is not. The Toshiba T1800 carries a retail price tag of $1,800 and can probably be bought "on the street" from mail-order discount stores for around $1,200. Toshiba's cellular modem, the T24D/X, retails for $259.

The latest developments in telecommunications technology are in the field of satellite navigation and communications. Comsat Maritime Services' C-Link system, coupled with the Trimble Galaxy Inmarsat-C/GPS system, is an example of a seagoing computer that becomes a fully integrated communications and navigation station capable of obtaining weatherfax, sending and receiving data, handling financial transactions, getting the latest news—everything one would expect from a computer at home or the office. (See Figures 4-3a and 4-3b for a photo and a diagram and description of how the Comsat and Galaxy Inmarsat systems are combined and installed on various-size vessels.)

Although the technology/cost curve for satellite communications is falling rapidly, the equipment is still not inexpensive by any assessment. The typical hardware installation, including notebook computer, Trimble Inmarsat interface, and antennae, ranges from $15,000 to $25,000, depending on specifications and manufacturer. Telecommunications time is currently billed at $5.50 per minute, or roughly the equivalent of the charge for overseas telephone calls. For that you have a fully functional, discrete telecommunications system, capable of sending and receiving all modes of communications—voice, data, and fax—with unlimited range. For a cruising yacht over 30 feet, or for a commercial vessel,

60 The Seagoing Computer

Figure 4-3a. Trimble Navigation integrates GPS and satellite telecommunications with a notebook PC and printer to create the ultimate space-age full-service navigation, computer, and telecommunications package. (Photo courtesy of Trimble Navigation)

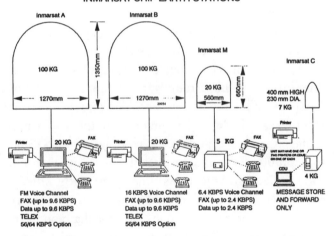

Figure 4-3b. This drawing illustrates the Inmarsat earth stations and their capability, associated equipment, and applicability to vessels of various sizes. (Illustration courtesy of Comsat Maritime Services)

Choosing and Linking Computer Peripherals   61

Type of Communication:	Inm-A Analog	Inm-B Digital	Mobile Link(**) Inm-M Digital	C-Link(**) Inm-C Digital Store & Fwd.
Applications:				
Voice	X	X	X	
Telex	X	X		X
Low Speed Data	X	X	X	X
Hi Speed Data	X	X		
Compressed Video	X	X		
Interactive Communications	X	X	X	
Polling	X	X	X	X
Broadcast/Group Calls	X	X	X	X
Paging	X	X	X	X
Vessel Tracking / Monitoring				X
SCADA				X

Type of Communication:	Inm-A Analog	Inm-B Digital	Mobile Link(**) Inm-M Digital	C-Link(**) Inm-C Digital Store & Fwd.
Antenna Type:	Stabilized Directional	Stabilized Directional	Stabilized Directional	Fixed Omni
Above Deck Equipment*:				
Size of Antenna (in)	54H x 54W	54H x 54W	24H x 24W	6H x 17W
Size of Antenna (cm)	140 x 140	140 x 140	61 x 61	15 x 44
Weight of Antenna (lb)	200	200	40	under 10
Below Deck Equipment*:				
Size (HxWxD in)	14 x 18 x 14	TBD	6 x 16 x 10	5 x 8 x 10
Size (HxWxD cm)	36 x 46 x 36	TBD	16 x 41 x 25	13 x 33 x 25
Weight (lb)	80	TBD	10	5
Capabilities - Voice:				
Type	FM	16kps	4.8kbs	N/A
Quality	Excellent	Excellent	Good	N/A
Capabilities - Data:				
Maximum Rate	56kbs	16kbs	2.4kbs	600bps
Fax Rate	FM	9.6kbs	2.4kps	TBD
Approximate End User Costs (US$):				
Terminal Price	$30-40k	$30-40k	$15-20k	$8-10k
Telephony/Min.	$7-10	TBD	$5.50	N/A
Telex/Min.	$4-5	TBD	N/A	$4-5
Data	$10/min.	TBD	TBD	$1.00/kbit
Availability	available	late 1992	mid-1992	available

* Exact size of antenna and below deck equipment varies with manufacturer, these are estimates and customers should confirm specs with each manufacturer.

Figure 4-4. This chart details the services, specifications, and costs of the mobile communications services provided by Comsat Maritime Services.

this capital outlay is not overly burdensome, given the added capabilities and safety afforded the crew. As with all technological innovations, you can look for the price of the systems to decrease over time. (See Figure 4-4 for a cost breakdown of components.)

A full description of the capabilities and features of the Galaxy Inmarsat-C and C-Link systems is presented in Chapter 8.

Computer Printers

In spite of much talk about the coming "paperless society," most mariners who work aboard will need or desire at some point to print various documents, letters, reports, weatherfax charts, and lists. So a printer is another virtual necessity for the seagoing computer office. Fortunately, printers have become much less expensive in the past few years, as well as smaller, faster, and better at producing good-quality type.

As recently as the mid-1980s, printers were big, heavy, slow, and expensive. The typical letter-quality printer weighed 30 pounds, cost $600 or more, and printed only about 30 to 60 characters per second, with a noise level comparable to that of the engine room of the Staten Island ferry. A one-armed secretary with a bad attitude could type faster. Today, a 4-pound ink-jet printer costing $200 to $300 can silently produce four pages of laser-quality text per minute, while taking up no more space on a nav table than a copy of *Chapman's Piloting* or a VHF radio transceiver. This is one more technological advance that makes a seagoing computer feasible.

There are myriad other printers, ranging from $250 to $300 dot-matrix printers that produce near-letter-quality type to state-of-the-art laser printers like the Hewlett-Packard Laserjet III, IIIp, and IV series. These laser printers are capable of producing flawless, almost-typeset-quality output, and they offer a wide range of type faces. They are the best printers for desktop publishers, newsletter editors, and graphic designers, but they are expensive, generally costing from $1,500 to $3,000. Both laser and dot-matrix printers take up more space than an average boatowner or commercial operator might like, and they can be heavy. The new ink- or bubble-jets represent the best overall choice in terms of quality output for the dollar, and of the space required at the nav table or workstation.

The Kodak Diconix, the Canon BJ10E Bubblejet, and the Toshiba ExpressWriter 201 are examples of printers that employ ink-jet print heads for high-quality printing of any document. Their ink supplies are contained in the print-head modules, eliminating the need for ink or toner cartridges or ribbons.

The Diconix 150 Plus by Kodak (Figure 4-5) is small (2 inches high by 11 inches wide by 7 inches deep), light (3 pounds), and can run on internal C-size batteries or a 9-VDC power supply. It can also be powered with 110 VAC, directly or through an inverter. The diminutive machine can print on single sheets or continuous forms and produces good print quality for both characters and graphics. Unlike the very expensive toner cartridges for laser printers, these ink-jet machines use small print heads that cost around $10 to $12 each. They're good for around 500 pages per print head. The printers are compatible with IBM PCs (or compatibles) and the Apple Macintosh line of computers.

Figure 4-5. The Kodak Diconix (left) and the Canon BJ10E Bubblejet (right) represent two of the newer, small-footprint portable ink-jet printers. (Photo courtesy of D.F. Crane Associates, Inc.)

The Canon BJ10E Bubblejet is slightly larger than the Diconix and weighs about the same. It can run on 110 VAC directly or through an inverter, or on an optional battery pack. It too can print on single sheets, using an optional sheet feeder. Its manufacturer claims good print quality on graphics and superior quality on characters. Its print heads are twice as expensive (about $25) as the Kodak Diconix printer's, and the printer is compatible only with IBM PCs and compatibles.

Either choice will work well at sea if your computer is an IBM PC or compatible. Only the Diconix will work with the Macs, however. The Canon prints characters in a variety of fonts, with better quality than a 24-pin printer, and is a better choice if you produce a lot of correspondence. The Diconix print quality is better than a

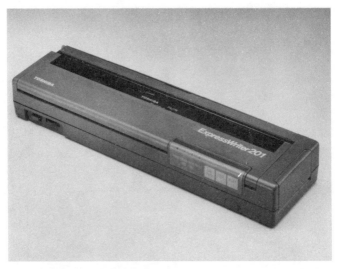

Figure 4-6. The Toshiba ExpressWriter 201 appears to be ideally suited to the restricted space available on most yachts, but its future is in doubt. (Photo courtesy of Toshiba America)

9-pin printer, requires less power, and uses less expensive cartridges. They print graphics about equally well.

As of this writing, the fate of the Toshiba ExpressWriter 201 (Figure 4-6), actually the smallest of these three printers intended primarily for laptop and notebook use, is in doubt, according to Toshiba America.

Compact Disc–CD-ROM

Possibly the most exciting development in the computer industry in the past few years is the adaptation of compact-disc technology as a storage and publishing medium for computer users. Compact discs (CDs), those small, highly polished disks that deliver such superb sound from your stereo system, are also capable of storing millions of bytes of data accessible by a special reader attached to your computer's motherboard. Instead of crowding your hard disk with tens of megabytes of data, or instead of using 3½-inch floppy disks, each capable of holding a mere 1.44 megabytes of data, you can now have all sixteen volumes of the Oxford English Dictionary, the history of the world, or all the navigation charts of the U.S. East Coast on a single CD. All of Joseph Conrad's novels will fit on a single CD—and I hope someone will eventually have the good sense to publish them on a CD. A company is now publishing the white pages of every telephone directory in the United States on two or three 2¼-inch-diameter CDs! The possibilities are mind-boggling.

For the mariner, the ability to access navigation charts, celestial and tide charts, and color graphic simulations of harbors and approaches will revolutionize navigation and nautical life in general. Figure 4-7 gives a listing of titles offered by just one company for the IBM PC and Mac. There are now hundreds of companies publishing CD-ROM titles, and the number is growing, with titles that include navigation charts and a plethora of celestial, GPS, SatNav, Loran, and other navigation aids. The marine trade press carries a rapidly growing array of products available from manufacturers and publishers in this area.

NEC is one of the world's largest producers of CD readers. Computer users can choose between an internally mounted CD reader that fits inside a CPU case on full-size PCs and Macs or an external CD reader, which is about the size of a paperback book and can be connected to the computer by means of a special parallel-interface adapter.

The NEC CDR-74 external CD-ROM reader sells for $725, and the required interface kits sell for $89 and $37 for the PC and Mac, respectively. These prices are all full retail. The hardware can be purchased for far less from discounters, and you can expect prices to fall as technology develops, and more manufacturers enter the field and competition increases.

CD *Express*—PC Version

Lucasfilm Games Loom—A wondrous game that returns you to the Age of the Great Guilds, where you'll solve the Mystery of the Disappearing Weavers. Magical graphics and enhanced audio add to the realism.

Figure 4-7. A sampling of software title descriptions.

Publish It!—Everything you need to quickly create your own newsletters, brochures and much more. Includes spell checker, thesaurus, clip-art images, drawing tools, and sample documents.

Great Cities of the World, Volume 2—A multimedia tour of ten of the world's most exciting cities! Discover the best hotels and the finest restaurants, and preview the most exciting attractions and entertainment.

Interactive Storytime—Three delightful children's stories that both entertain and enhance reading development and language skills. Listen for pronunciation as a storyteller reads aloud, or print the illustrations in coloring book form.

Total Baseball—A fascinating in-depth history and complete statistics of the game of baseball, players and team profiles, photos and audio recordings of many of the sports's finest moments. The ultimate trivia source for baseball lovers!

The Family Doctor—A practical, extensive medical guide offering anwers to nearly 1,500 commonly asked health questions. Includes national support group listings, educational resources, and data on over 1,600 prescription drugs.

GeoWorks CD Manager—Easy-to-use organizer provides an audio control panel for playing your music CDs, plus easy access to CD Express CD-ROM applications. Also includes appointment calendar, address book, Tetris, solitaire, and more.

Figure 4-7, continued. A sampling of software title descriptions.

The Software Toolworks Reference Library—The award-winning new standard for home computer libraries. Includes Webster's New World Dictionary, thesaurus, guide to concise writing, and other essential resources.

Bureau Development Inc, Best of the Bureau—A collection of works of literature and history, selected from the Bureau Development library. Featuring Monarch Notes, the complete Dickens, Biographies, U.S. and World History.

Ultima VI, the False Prophet/Wing Commander—Two acclaimed adventure games on one CD. Lead a team of knights through a perilous medieval world in the False Prophet; and soar through space on Wing Commander, the futuristic flight simulator.

Just Grandma and Me—Characters in this highly interactive storybook come to life through brilliant animation. Almost every object on the full-color screen has an audio and video action, activated by a simple mouse click as the storyteller reads aloud.

Great Wonders of the World, Volume 1—Take a multimedia guided tour of the Man-Made Wonders of the World. Each spectacular location is brought to life through motion video, photographs and audio. With extensive travel information, from hotels to restaurants to attractions.

Figure 4-7, continued. A sampling of software title descriptions.

Sherlock Holmes, Consulting Detective—A fascinating multimedia adventure game in which the world's greatest detective helps you solve three puzzling crimes. Watch and listen for clues as Holmes describes each case and questions suspects during extensive motion video scenes.

Color It!—The only limit is your imagination in this award winning, easy-to-use image editing and paint program. Spectacular 32-bit technology lets you create up to 16 million colors, and powerful image enhancement tools let you use a variety of truly special effects.

The New Grolier Multimedia Encyclopedia—A remarkable multimedia research tool with exciting motion video and sound sequences, 33,000 articles and more than 3,000 photos and illustrations. Selected Product of the Year by the Optical Publishing Association.

The Software Toolworks U.S. Atlas—See the U.S.A. with full color, highly detailed maps. Analyze over 1,000 statistical maps covering over 200 topics. You can even create new topics and add data to the permanent database.

Figure 4-7, continued. A sampling of software title descriptions.

Communications Utilities

Modem/Terminal Software

One of the keys to having a truly practical computer workstation aboard your boat is the ability to communicate with other computers to send and receive documents, faxes, and electronic mail—in other words, to fully link your boat with the world ashore. As mentioned in the last chapter, this is possible through telecommunications using a modem hooked up to a computer. Communications that seemed farfetched a few years ago are possible today, and at reasonable cost. This chapter deals with features to look for in modem software programs, electronic on-line information services, and weatherfax.

A modem requires one or more software programs to function. A terminal program enables a computer to access telephone lines to send and receive electronic signals. A second program is needed in order to send and receive faxes.

The Terminal Programs

Terminal, modem, and telecommunications programs—all of which are practically interchangeable terms—are software applications designed to allow computers to communicate with modems, thus enabling you to communicate via the vast electronic world. That electronic link might be only between you and your office. It could be between your boat and anywhere in the world, using a global array of telecommunications and navigation satellites set up to guide you and allow you to communicate from wherever you are. Or it might be used to access a rapidly expanding universe of information and utilities. A global community of computers will one day connect us all, but whatever your intended level of involvement within the electronic global community, you'll need a modem program.

If you're an IBM-PC owner, you'll have been furnished—or at least enticed by the Microsoft Corporation to buy—its Windows Graphic User Interface (GUI) Operating System, hereinafter and universally referred to as "Windows." Windows has its own built-in terminal program, but it is not one of the better Windows features (it's cumbersome and very short on features), and you might elect to install one of the many MS-DOS–based terminal programs, or the many very sophisticated telecom programs now written especially for use with the Windows environment.

If you're a Macintosh user and fan (the two seem to be synonymous), you'll be using either a version of System 7, the Mac's latest proprietary operating system, another GUI at least as good as Windows for the IBM PC and compatibles, or perhaps a Mac version of Windows itself. Either way, a terminal program is standard equipment for Macs too.

There are various commercial, "shareware," and "freeware" telecommunications programs available for IBM PCs and compatibles and Macs. Commercial programs, such as DC/Crosstalk, Procomm Plus (for Windows or Mac), and Hayes Smartcomm, cost from $50 to $500. Each has different versions and capabilities. Most MS-DOS terminal programs allow the user to operate the program from within the Windows environment. Many shareware programs, such as Unicom and Telix, can be downloaded from BBSs or commercial utilities like CompuServe, Delphi, GEnie, Prodigy, and National Videotext. The royalties for this shareware, if you decide to use it beyond an initial test period, range from $39 to $69. These programs are full-featured and professionally written and documented, and they will suffice in place of the best of the expensive commercial programs.

Fax Programs

If you are installing a commercial or shareware telecommunications program, you'll also want fax send-and-receive capabilities. To do that, you'll need send/receive fax communications software, which is often bundled along with the hardware. These programs are quite complex, and shareware-software programmers haven't yet really addressed this area.

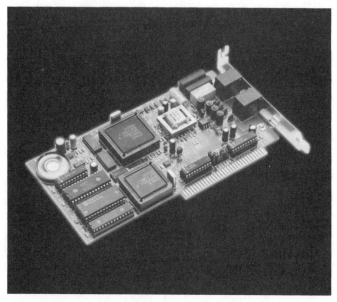

Figure 5-1. The Supra Fax Modem, shown here in its internal version.

An excellent package is the Supra Fax Modem V.32bis (Figure 5-1), which sells for about $400 and is available as either an internal or external unit. It comes complete with modem, cable, and excellent Winfax Pro software by Delrina Technologies. The Supra is capable of data-transmission speeds of up to 57,500 bps, one of the fastest for the price, and the Winfax software allows a PC user to create, send, and receive professional-looking faxes right from the Windows "desktop." A document created in, say, Word for Windows, Word 5.0 for Mac, or virtually any of today's popular word-processing/desktop-publishing programs, even with a variety of fonts, graphics, and tables, can be sent using the Winfax program and the Fax Modem. In receiving a fax, Winfax Pro uses optical character recognition (OCR) to "read" virtually any typed or typeset text, graphics, and even handwritten margin notes and signatures.

Again, any of this new-technology software written for IBM and compatibles is also available for Mac. Since the Mac System 7 operating system can read either Mac or MS-DOS software, most programs are interchangeable between systems. However, in most cases when buying commercial software, you'll find that the computer system for which the product was written is usually indicated on the package.

The professional person who lives and works aboard, as well as the mariner who cruises year 'round, can now send any document from an onboard computer terminal, through a telecommunications program, and over telephone lines or via satellite telecommunications simply, inexpensively, and at the blinding rate of 50,000 digital computer bits of data per second.

On-line Information Systems and Databases

Beyond the creation, editing, transmission, and management of documents, there is an entire world of information, news, reference data, entertainment, and services available to the seagoing computer user via on-line information services. These range from simple bulletin-board systems that target a particular audience or interest group to highly sophisticated on-line information systems like Delphi, Dow Jones News Retrieval, GEnie, National Videotext, and Prodigy. There are also huge databases, such as Lockheed California Company's

DIALOG, a mind-boggling array of information resources organized under a single proprietary commercial-access system, and Mead Data Central's NEXIS and LEXIS databases for the legal profession.

Access to this impressive cornucopia of information resources is a subscription to one or more of these on-line information services. For a few dollars per month, a computer user with a modem can log on to and choose from thousands of services and files, from encyclopedia-type info and banking and shopping services to instant access to news and financial wires and electronic-mail messages. In fact, virtually any topic you can imagine has its electronically accessible interest group somewhere on-line.

When subscribing to any one of the big on-line information services, like CompuServe or Delphi, a subscriber is given a comprehensive manual on how to use the system as well as a thorough guide and index to its on-line resources. Most of the principal on-line services, which are listed in Appendix B, also publish monthly newsletters or magazines to further educate and update their subscribers as to what's new and available on-line.

On-line services can make work a whole lot easier and faster. To illustrate, I once was given an assignment to write an article about an area of law with which I was very familiar but that involved substantial collateral research in medicine, law enforcement, science, and sociology. Five years ago, I would have spent a week of eight-hour days in the public library, and possibly would have taken trips to area university, medical, and law libraries as well.

Not today. By logging on to the Delphi on-line information service, I was able to access everything I needed in about an hour. In writing this long, detailed article, I cited sources as diverse as *Grolier's Encyclopedia*; *The New England Journal of Medicine*; Rutgers University; the Georgetown, Harvard, and University of Colorado law libraries; the Federal Bureau of Alcohol, Tobacco, and Firearms; the Drug Enforcement Agency; the Federal Aviation Administration; the Federal Bureau of Investigation and other elements of the U.S. Department of Justice; several state supreme courts; and the U.S. Supreme Court. Furthermore, I did it all while relaxing in the main salon of my yacht listening to classical music, the ebb and flow of tides and harbor traffic, the plaint of seabirds, and an occasional ship's bell.

Since I was logged on to the service through a local telephone number, there were no toll charges. Due to a special plan at Delphi, I was able to accomplish all of this research for the astonishingly low cost of one dollar. Delphi has a plan that enables a subscriber to purchase twenty hours of on-line time per month for $20, and all of the major on-line services have a similar plan. (Many of the noncommercial BBSs cost nothing at all and have thousands of downloadable information files and executable programs.)

Via an interlink through Delphi called "The Internet," a subscriber can connect to virtually every major electronic information resource in the world, including nearly every university library and research laboratory, U.S. and foreign government and military agencies such as NASA and the Department of Defense, military bases and other installations, small and large corporations, and practically all of the other commercial on-line systems.

With the information age upon us, the "electronic cabin" is a reality. Who needs an office ashore?

Weatherfax

Knowledge about weather, combined with accurate forecasting, is important for safety aboard any boat, whether you are navigating on large lakes or along a coast, or making long passages. With an onboard fax receiver, you can obtain timely weather reports, forecasts, synopses, advisories, and warnings from the National Weather Service (NWS) at any time of the day or night.

According to Accu-Weather, Inc., a commercial provider of weather facsimile charts and other weather-reporting and -forecasting products for marine and aviation users, a mariner can obtain up-to-the-minute weather information from a choice of thousands of weather charts and other weather media. Called DIFAX, these charts include Significant Weather Prognoses, Radar Summary Charts, Weather Depiction Charts, Upper-air Analyses, and Surface Analyses. You can also obtain significant weather watches posted by NOAA (National Oceanic and Atmospheric Administration), the National Weather Service, all marine forecasts and current reports, climatology graphs, and other meteorological information at sea, just as you would from the National Weather Service office ashore.

Costs for these services are on a per-item basis, and the company says they charge no sign-up or initiation fees to get on-line. Accu-Weather does charge a $49 annual fee for faxed notification of official NWS weather warnings, however, plus a small fee of $1.99 per notice issued to your onboard fax machine.

These notices include severe-weather advisories, but they can also be tailored to include other watches and warnings posted by NOAA and NWS: tornado watches;

Communications Utilities 79

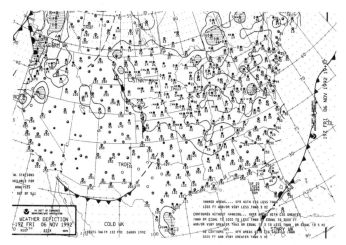

Figure 5-2. NOAA weather depiction chart, as recieved by fax. (Courtesy Accu-Weather, Inc.)

severe thunderstorm, hurricane, tropical storm, flooding, and high-tide warnings; and high-wind, heavy-rain, snow, ice, and dense-fog conditions. Notices are received from all of the meteorological agencies, including the National Severe Storms Forecast Center, the National Hurricane Center, and local National Weather Service offices.

Aboard today's computer-equipped vessel, it would be hard to justify doing without the added safety feature of facsimile weather warnings and reports. Fortunately, few will need to, thanks to the emergence of inexpensive fax capability for any computer installation.

Software for the Seagoing Computer

Once you've installed your computer system on your boat, you are ready to configure it to do the work you want to do.

The basic tasks of a seagoing computer usually involve word processing, file management, mathematical and accounting spreadsheets and databases, and yacht-management functions such as budgeting and stores, maintenance, and inventory tracking. We'll discuss some of the more popular programs available to accomplish these tasks.

Word Processing and Desktop Publishing

The revolution in word-processing programs that has taken place over the past three or four years is truly amazing. Not long ago, word processing meant exercising an incomprehensible set of non-intuitive keyboard commands to manipulate green letters on a black screen. Formatting was a nightmare completely understood only by the authors of the word-processor machine code. What you saw on the screen was, well, not necessarily what would be printed on the page. Today we have Mac and IBM PCs with Windows, and word-processing software with true what-you-see-is-what-you-get features that make creating virtually any kind of document as easy as turning on the machine and typing. If you can type, even by the timeless hunt-and-peck method, you can create any letter, report, novel, newsletter, memorandum, legal brief, screenplay, or fax the mind can imagine.

Word-processing software on the market now has all the capabilities the most sophisticated page-layout programs, known as desktop-publishing systems, had five years ago. Two of the top programs, Microsoft Word and Lotus Development Corp.'s Ami Pro, are fully capable of on-screen typesetting, and almost-typeset-quality printing using laser printers. Both programs sell for about $495 at retail. However, as part of a trend that is virtually industry-wide, most computer manufacturers bundle one or the other of these wonderful word-processing programs with every computer. Since even the mail-order deep-discounters bundle software for word processing, spreadsheets, and telecommunications with computer systems, the only reason to buy a new word-processing program is if you are installing an older computer on board that doesn't have a state-of-the-art word-processing program.

No matter what you are doing, whether preparing a business or personal letter, writing a weekly or monthly newsletter, or preparing a book manuscript, you can and of course should, turn out high-quality manuscript and graphics easily and inexpensively. In fact, these capabilities lend themselves to small businesses such as newsletter production and book publishing that would have required much time and work ashore only a few years ago.

Word for Windows, Ami Pro, and Wordstar 6.0 are examples of word-processing programs that can create text in a wide variety and size of typefaces; cut, paste, and otherwise manipulate text; format columns of text; and insert graphics and spreadsheets—all now considered standard features of a first-rate word-processing program. Using Word for Windows, for example, a writer can choose from an almost infinite number of type sizes and hundreds of optional typefaces. Document templates are available for form letters, business and personal letters, reports, memoranda, legal documents, academic dissertations, magazine and journal articles, and other specific applications that eliminate worry about formats as well as the need to purchase expensive preprinted letterheads and envelopes. Text can be "snaked" into multiple columns on a page for newsletters and articles.

The high-end personal-publishing programs are also capable of checking grammar, spelling, and punctuation, counting words and paragraphs, and evaluating writing style. Although these stand-alone programs are primarily for text management, final documents can be integrated with spreadsheet, database, desktop-publishing, and telecommunications programs.

Although there are very sophisticated desktop-publishing programs, such as the highly acclaimed Aldus Pagemaker and others, unless you are looking to create color, magazine-quality finished copy, the aforementioned state-of-the-art word-processing programs are quite sufficient.

Multipurpose Software Programs

A number of very capable programs that combine text editing, spreadsheet, database, and telecommunications functions in one program are also available. One example of such a program is Microsoft Works, the electronic equivalent of the Swiss Army knife. This versatile program comes in MS-DOS, Windows, and Macintosh versions, and incorporates a reasonably powerful word-processing program, adequate spreadsheet and database programs, and a very adequate terminal or telecommunications program. Note that I have qualified these descriptions. These programs should not be compared with far more powerful and feature-laden stand-alone programs, because programs designed for versatility and simplicity often require some compromises.

For the average user, however, multipurpose programs offer an inexpensive, easy-to-use package of the computer software used most often. Although you might not wish to design a newsletter or brochure with Microsoft Works, there's no reason not to compose the text for one using its word-processing capabilities. The program takes up very little hard-disk space and allows the user to switch quickly from text to database to spreadsheet to terminal program from a single menu. Its word-processing program has a spelling and grammar checker, and its text-formatting features are perfectly adequate for most documents. The spreadsheet and database programs are adequate for the average small-business user or home-accounting routines, and the terminal program, while not as sophisticated as the leaders such as Bronco or Crosstalk, is the essence of simplicity. The program has powerful features such as log-on scripting, a program to automate the sometimes complicated process of connecting to your favorite remote-access on-line systems. It incorporates a telephone directory, calculator, calendar, clock, Rolodex, and file manager. I use Works on board because I often use a laptop computer, and the Works program is a fine integrated program to use on a laptop.

Text and other files generated with Works can be opened and edited directly by Microsoft Word; the files can be used in other software programs after being run through one of the several conversion programs available. Thus, files created on board on a laptop or notebook computer, using the small, efficient Works, can be worked on in a stand-alone program you might have on your desktop computer or workstation at home or at the office. Files can be swapped between Works and Word, as well as between IBM compatibles and Macs. Menus, protocols, and commands in Works and the entire range of Windows products are very similar. Microsoft Works and Microsoft Works for Windows sell for about $149 retail, and sometimes even less at discounters' "street" prices.

Spreadsheet and Accounting Software

Lotus 1-2-3, Borland Quattro Pro, and Microsoft Excel are good examples of what's available in state-of-the art spreadsheet and accounting programs. These integrated programs can generate not only numbers in columns and rows but text and other data as well. They can also use data from word-processing, database, and graphics programs.

Sophisticated spreadsheet programs like Lotus Development Corp.'s 1-2-3 and Microsoft's Excel are overkill, however, if what you need is simply to keep track of a yacht budget or check register. In that event, smaller, simpler programs such as Microsoft Works are more than adequate. The big programs also use 10 to 20 MB of hard-disk space and are intended for accountants and other business professionals who need a powerful spreadsheet and accounting program. Prices for these programs are generally between $300 and $400 in mail-order catalogs, although one of them is often bundled with new computer purchases. These software titles, or their equivalents, are available for IBM and IBM-compatibles, as well as for Macs.

Yacht-Management Software

There are some highly sophisticated and industry-specific software programs on the market today designed for marine applications such as engineering, vessel design, marine architecture, and commercial-fleet management. These, however, are beyond the scope of this book. A number of programs have also been developed to aid yacht and smaller-boat owners, liveaboards, cruisers, or racers. These programs are advertised in yacht, sailing, and powerboat magazines, as well as in mail-order catalogs.

Bluewater Yacht Manager, Version 2.0

Bluewater Yacht Manager is a program written to aid anyone planning cruises or living or working aboard a boat. The Bluewater Yacht Manager automatically calculates costs of projects, improvements, and repairs. It helps make decisions about outfitting or refitting, and can analyze and prioritize maintenance projects for cost efficiency, safety, and budgeting. The software helps create and maintain lists of food and boat stores and equipment, providing reminders when it is time to restock. It inventories and indexes other data such as charts, has a passage-planning module, and keeps track of crew information such as vital statistics and medical and immunization records and schedules. It is also used to keep a boat's complete maintenance and sailing logs.

Bluewater Yacht Manager can be operated on any full-size or portable IBM, IBM-compatible, or Mac with the appropriate version, and can be used with any printer or screen display. It sells for approximately $129. It is produced and sold by Shelter Island Press of San Diego. It is also available from mail-order firms such as D.F. Crane Associates.

Passagemaker

This program is a system intended to enable the yachtsman to plan for extended cruising. The program tracks the trip and aids in keeping the daily log, as well as crew information, visa and permit information, chart inventories, and food and supply lists. It also features lists for equipment and parts, as well as a schedule for repairs. It has a full-featured budget and expense module and will generate reports pertaining to the business use of the vessel for business-accounting and tax purposes.

This program will run on any IBM or IBM-compatible computer with one or more floppy diskette drives and running DOS 2.1 or greater. It sells for $199 and is available from its developer, MicroMarine, and mail-order catalogs.

Macintosh Datalocker

Datalocker is a software program developed for the Mac. It uses the Macintosh proprietary system known as Hypercard to provide a comprehensive vessel-management program. Datalocker can classify tasks by individual boat systems, such as engine, fuel, water, lighting, and steering, and can generate lists for planning purposes. Vessel-specific engineering and systems data is recorded on templates, facilitating repairs and maintenance while underway. The program will also output user-defined reports covering all areas of yacht management and operation. Datalocker, which sells for $495, is available through mail-order catalogs.

Public-Domain Software for Marine Use

There is a large—and growing—number of computer programs designed for the marine user available on commercial and private bulletin-board or on-line services that are free, or nearly so, by simply calling up and downloading them to your computer.

Several hundred such programs, covering yacht management, cruising, racing, and liveaboards, are available from such information sources as Delphi, CompuServe, and the Sail BBS. The telephone numbers and other information for these are provided in Appendix B, "Telecommunications and On-line Services."

7

Maintenance and Troubleshooting of Computers and Peripherals

The compact, well-designed, and precisely engineered computers currently on the market usually, and fortunately for onboard users, don't require much in the way of maintenance. This book, for example, is being written in part on an IBM compatible that was taken out of the box, set up, plugged in, and has been running continuously since 1990. It is operated in an office where the temperature is more or less constant, the power source is fairly clean and steady, the machine is stationary, and the workspace is kept relatively free of dust, dirt, and moisture. Needless to say, these conditions are rare aboard a boat, so there arise a few specialized maintenance and troubleshooting considerations when operating in that unique environment.

Because laptop computers are small, sealed, and tightly engineered, they are far less likely than full-size desktop computers to fall victim to the particular environmental hazards found at sea. Laptops need to be kept clean and dry, and we'll discuss a few problems specific to seagoing laptops and notebooks in the troubleshooting segment of this chapter.

The primary enemies of conventional computers on board are moisture, corrosion, temperature fluctuations, motion, power faults, sea air, and vulnerability to physical damage from a wide variety of shipboard activity. These variables are listed in no particular order; they all can come into play.

It goes without saying that a computer system drenched by seawater pouring in from an open hatch is going to suffer. It's no different from a radio, stereo, or nav receiver in that respect. The same precautions taken to avoid dousing these common ship's accessories should also be taken to avoid water damage to a computer. Installing the system away from hatchways and ports is a good idea, and keeping such disasters in mind while underway will help to avoid electronic catastrophe.

A more likely cause of moisture damage—and one far more insidious and difficult to prevent—is the effect of the marine environment's moisture-laden air and general high humidity. Early attempts to protect onboard computers included sealing components and processor cases with cellophane, plastic, aluminum foil, or electrical contact aerosol sprays, etc. But the machines weren't engineered for such do-it-yourself measures, and the cure was usually worse than the disease.

Instead, keep the computer area as dry you can. Use a dehumidifier aboard your boat, if at all possible, and avoid leaving the boat cabin closed up with the computer inside for long periods of time, particularly in damp or warm weather.

Food and beverages have an uncanny way of ending up inside keyboards and CPU cases, so inform everyone about the need to stay away from the computer while eating or drinking. I once diagnosed a mysterious disk-drive failure on a brand-new computer installed on a yacht that was home to a young couple with a small child. We discovered, to our amazement and barely concealed mirth, that a graham cracker is almost exactly the same thickness as a 5¼-inch floppy disk and will slip easily through a disk-drive door.

Although dust is not usually a problem in a marine setting, be mindful of delicate computer components while sanding, scraping, cleaning, or undertaking any other dust-generating activity on board.

Extreme temperature fluctuations can also cause problems. Such internal computer components as hard-disk drives and CD-ROM readers are somewhat temperature-sensitive. Try to avoid exposing a computer system to long periods of extreme heat or cold, and to rapid changes in temperature.

Motion, unavoidable on a boat, is less of a concern for today's more sophisticated internal components. But the rotating drums of CD-ROM readers and hard-disk drives are subject to increased wear from motion in a seaway or rough mooring conditions, so avoid using the computer in such conditions if possible.

Of course, collision with objects propelled by heavy seas, or the computer itself falling from its mount in these conditions, would obviously cause major damage. Take extra precautions to secure your system when these conditions are anticipated, and make your computer system one of the first items secured during any unexpected encounter with rough weather.

Sea air, no matter how balmy and tranquil, can seriously corrode wire, solder, and metal on printed circuit boards in a computer system. This is nothing new, of course, nor is the computer system necessarily any more vulnerable to corrosion than radio gear, electronic nav equipment, and other instruments. Constant vigilance in covering, protecting, and cleaning a computer is a requirement in all but the most arid atmospheric conditions. Tight laptops and portables have an advantage here; however, I've found that a twice-yearly dismantling and thorough cleaning of computer components, either by you or a computer shop, generally results in good overall performance and freedom from corrosion-induced failures. When read/write errors, hard-disk crashes, keyboard errors, or lockups not caused by software occur, suspect the influence of that marvelous sea air we love so well. Shut the system down and take the CPU ashore for a thorough inspection, cleaning, and, if necessary, replacement of the affected components.

A few words are appropriate here on a much-debated computer controversy: When, if ever, should a computer be shut off? Despite good arguments about energy conservation and the number of hours of expected service life of computers and components, my advice is: Don't turn it off. One of my computers has run continuously for four years, and has never failed. Obviously a seagoing machine should be turned off when the boat is left for any period exceeding overnight, and at most times when underway and not in use, due to electrical power considerations. However, the best rule is to turn the machine on and off as little as possible. And when you do turn the machine off, particularly to cold-boot the system (it's better to use the "reset" switch), wait at least 30–40 seconds before turning it back on. This will prevent residual electric current and/or static electricity in the system from "spiking," or causing an over-voltage fault.

Power faults are the particular menace of any computer that doesn't operate off a stable source of electrical current. Ship's power, batteries, gen-sets and alternators, shore-power facilities—all are subject to a lower standard of consistency than most commercial and/or residential power sources ashore. Problems may range from under- or over-voltage output that exceeds computer limitations and results in malfunction, to significant power fluctuations that can, in many instances, crash hard-disk drives and CPU controllers. Quality surge suppressors, discussed in Chapter 3, are an absolute requirement of the seagoing computer. Additional protections, such as continuous power sources and an alternate power supply, are desirable but, frankly, impractical, due to cost, weight, availability, and space considerations.

A number of precautions can ensure the best possible power source for the computer system. First, a good ship's voltage/amperage indicator, properly tested and calibrated, should be part of the boat's power-station instrumentation. Monitor this instrument when attaching shore power or switching to ship's batteries, alternators, or other onboard power sources, or whenever there appears to be an electrical-related problem with the computer installation. A great deal of diagnostic work can be done right there at the voltage meter.

This same guideline applies to the protection and diagnosis of all other electrical components aboard. It is a good idea to use a well-written manual, such as Kevin and Nan Jeffrey's *Boatowner's Energy Planner* (International Marine, 1991), as a comprehensive guide to your boat's electrical systems.

Beyond protecting the system physically and providing it with a clean and consistent supply of electricity, the machine's most critical—and vulnerable—component, its hard-disk drive, must be maintained with care. Since this is where all the data, software programs, operating code, and files are kept, it is of vital importance. As one of the computer's few moving parts, it is subject to damage and deterioration throughout its life.

Internal wear and/or deterioration is one of the biggest causes of disk damage and failure. To combat this, become familiar with some very useful and important utilities. One is part of your computer-operating system, such as MS-DOS or Apple System 7; the other is a commercial product, such as Norton Utilities, designed to analyze, diagnose, adjust, maintain, or, if necessary, recover your hard-disk drive after a crash.

The IBM PC, equipped with MS-DOS, or the Macintosh, equipped with Apple System 7, has a program, such as MS-DOS "Chkdsk," or the equivalent Mac diagnostic system available on its main menu, designed to check the status of a hard or floppy disk. The program reports on the amount of space available, and the amount being used by system and program files and user files. It also indicates the number of "allocation units," or places to store data, on a disk. When a "switch" is invoked, such as typing "/f" after the program execution command "Chkdsk" (i.e., "Chkdsk /f"),

DOS will not only check the disk attributes but "fix" any problems it detects with the allocation of data on the disk, the so-called file-allocation table (often the first place disk errors crop up). A word of technical explanation is in order here.

When data is stored on the disk—indeed, each time a file or a piece of information is called up by the user, read, edited, and then "saved" to the disk—the read/write head physically "reads" the information into the computer's Random-Access Memory (RAM) and displays it on your screen. When you are done with it, the head "writes" this data back onto the hard disk. Each time this is done, bits of data get scattered across the surface of the disk, like scraps of paper on your desk top, until the disk gets cluttered, or, in computer terms, "fragmented."

Eventually this cluttered condition begins to cause trouble. The solution is either to reformat the disk completely, which would result in erasing all the data, or to "unfragment" the disk with a program like Norton Utilities or PC Tools. There are many such programs, all of which clean house and perform a variety of other diagnostic and preventive-maintenance functions.

Programs like the Norton Utilities Speed Disk painstakingly scan the actual surface of the disk, meticulously inspecting each "room," or partition of the disk, and rearranging everything in it for efficiency. They don't stop until every microscopic area on the disk surface has been thoroughly attended to. This process may take hours, depending on the level of "pattern testing," or thoroughness, the user desires.

When all is tested, Norton will have dusted every square micromillimeter of the disk surface and arranged all your precious data into the least amount of disk space possible, leaving not a scintilla of "fragmented" bits lying around anywhere. The result is faster, more efficient disk access; quicker reading and writing to disk; less likelihood of lost or damaged data or read/write error; and, usually, more available disk space. Norton will also report on the amount of disk space in use versus the amount available, and the relative speed and efficiency of your CPU and hard-disk drive compared to current industry standards.

   The Norton Utilities and PC Tools also incorporate programs to diagnose an ailing disk, recover data in the event of a dreaded "crash" (which, mercifully, we rarely experience nowadays), and even reconstruct programs, files, and data lost during an inadvertent erasure of the disk. These features make such programs a near necessity for the serious user. They are also the leading system for detecting and eradicating the notorious computer "viruses" that have had so much adverse publicity in the last couple of years.

Common Computer Maladies and How to Correct Them

Everyone has heard stories about the, er, person who turns the computer switch "on" and, when nothing happens, calls the repair person, who comes to the site, finds the power cord unplugged, puts it in the wall, and presents a bill for $60.

   Beyond this kind of mere human error, a number of very common problems plague computer users regularly. Sometimes experience, or the lack thereof, will be a major factor in how easily—and quickly—the machine is back up and running. Frequently, the awareness of common computer faults will lead you to the light.

If a computer simply will not start up—no lights, no sound, no screen—it is literally not getting power. This could be something as simple as a power cord not plugged into an available source of 110 VAC, but we'll assume that something of a more serious nature, either within or without the computer, lies at the root of the problem.

If you are sitting in an office with a dead computer, but lights, stereo, and everything else is functioning fine, you might be tempted to rule out the source of electricity as the problem. Not so on a boat. Depending on the configuration of the electrical system, the computer and its associated peripherals and accessories might be connected to its own circuit. Or it might be on an electrical "bus" or subsystem with the ship's other electronics, but not with the lights, stereo, etc.

Start at the electrical circuit panel and see if there is actual voltage to the system from batteries, generator/alternators, or shore power. If the power is available to the circuit the computer is on, and you are able to trace the "juice" from the source to the back end of the computer's CPU, then it is possible that the machine's own internal power supply, the component that directs power from the cord to the computer's hundreds of tiny circuits, is faulty. Unless it is protected by a fuse (which is doubtful in any newer machine), the power-supply unit may need to be replaced. If you're handy and have a circuit tester, you can purchase a new power supply for between $69 and $160; if you're not, you can take the CPU to the repair shop.

It's also a good idea to determine why the power supply failed. This is not an item that gives up easily, and an external factor, such as a voltage spike or over-voltage surge, was probably the cause. Fix both the cause and the problem with the power supply, and double-check the adequacy of surge protection. Two of the most common causes of inadequate surge protection are a blown fuse on the surge suppressor, or having too many machines, radios, stereos, computers, printers, etc., plugged into the surge protector and exceeding its voltage limit.

"Boot" Trouble

The computer turns on and runs fine, but when you attempt to load programs, nothing happens. This is most likely a disk problem. It happens to every computer user eventually, though better hard-disk drives, fewer moving parts, and lower temperatures inside the CPU, among other reasons, have reduced the frequency. But it can happen, and to recover, you'll need a program like Norton Utilities, with its Disk Doctor, available in either IBM PC or Mac versions.

Norton Disk Doctor (NDD) will analyze your disk and tell you what's the matter. If the disk has "crashed" (a sort of catch-all term for anything between "failed" and "became so hopelessly fragmented that the read/write head couldn't make sense of it"), NDD will restore it to the extent possible. It will recover files, adjust file-allocation tables, or unfragment the disk. The program can't, of course, physically repair a damaged or worn-out disk, but it can tell you if that is the problem, or if it lies in the way the machine reads (or fails to read) the data on the disk. If the only problem is the most likely one—read/write trouble—Norton Disk Doctor will have you back in business very soon.

Another reason a computer will run mechanically but won't run software is a problem in the automatic execution batch file (known as the Autoexec.bat file) or the system configuration file (Config.sys file). Both these files must function properly in order to load and run the machine's operating system and other software programs. If you have recently installed new software, or reinstalled old software, suspect this problem.

Installing and/or setting up software usually causes the software-installation programs to alter one or both of these critical boot files. It doesn't take much of a bug in the syntax of these two "boss" programs to create havoc, particularly at start-up, because it is these two that tell every other program, including the computer's operating-system program, when to come to attention and whom to salute. If you can obtain a system-command prompt (c:>) but not the start of other programs, you probably have a Config.sys or Autoexec.bat problem. The solution is to issue commands to read these two vital programs, one at a time, and repair them if necessary. While a thorough discussion of every item these programs might contain is beyond our scope, a basic example of each will be given here so that you can go back to basics, get the computer booted up and running, and then reload additional software programs.

To view the two start-up files one at a time, type "type" (without the quotation marks) followed by the file name (i.e., Autoexec.bat) at the c:> prompt.

Here is a (very) basic Config.sys file:

DEVICE=C
FILES=30
BUFFERS=30

And here is a (very) basic Autoexec.bat file:

PATH=C:\;C:\DOS;C:\WINDOWS
PROMPT pg
SET TEMP=C:\...\TEMP

Note: Include WINDOWS in the PATH statement only if the Windows Operating System is used.

Note that I have emphasized "basic," because these programs are sufficient only to boot up the machine and, say, tell the system where to find the Windows program, if you're using it. You will not be able to call up other application programs without entering their directory unless they are included in the PATH statement, because the computer will not know where to look for them on the disk. They can be added to the PATH statement, whether by editing the file or reloading the software programs from their original floppy disks. Be careful with reloading though, to avoid a repeat of the original problem. It is best to load new, or previously used, programs one at a time to check for software-induced errors.

Software Errors

There are hundreds, maybe thousands, of possible software-error messages that might (but, again, rarely will) appear on your computer screen. The most common ones, like Windows's dreaded "Unrecoverable

Application Error," occur when the system doesn't have sufficient resources, like memory, to do its job.

Error messages are usually not as bad as they sound, although they can predict dire consequences, like hard-disk failure, loss of data, unrecoverable files, etc. When one of these messages appear, consult your computer-operations manual, as well as the user's manual(s) for the software program in use. These manuals will list each possible error message, along with the most likely solutions, in an appendix or chapter devoted to the subject. Many error messages contain their own explanation and the means of solving the problem. Windows and other Microsoft programs, and most Macintosh hardware and software, are thus self-diagnostic and often self-correcting.

Most of today's computers have vastly more sophisticated onboard diagnostic programs than those of even a few years ago. Likewise, most up-to-date software programs include interactive help files and hypertext explanations of what's going on inside them, and instructions on how to cure their ills. It is relatively rare to have to haul the box down to the repair shop anymore.

Moreover, almost all major hardware and software manufacturers have vast resources dedicated to customer service and support, with skilled technicians on the other end of (usually) toll-free hotlines to help you through the most vexing computer crises.

One exception is Microsoft Corporation. Though the technicians on the support lines are very knowledgeable and helpful, a long-distance call is required to reach them, unless you just happen to be in Seattle, and interminable periods are sometimes spent on "hold" waiting for a Microsoft representative to come on the line. Still, this beats taking the machine or the software program to the shop for service or repair.

Careful attention to installation and setup instructions, thorough planning of the onboard electrical- and computer-systems components, and a methodical approach to diagnosing trouble when it does occur can ensure years of relatively trouble-free use from your seagoing computer.

8

Computerizing the Helm and Navigation Tasks

The Marine Navigation and Communications Interlink

You're twelve hundred miles west of Hawaii, somewhere between Honolulu and the Marshall Islands. You awake as the sun rises dramatically astern and shines down into your pilot berth. The smell of the sea and brewing coffee fills the air, so you get up and swing over to the galley and pour a cup. Walk over to the chart table and tap the space bar of the laptop computer, and you are given a capsule briefing of what occurred during the six hours you were peacefully asleep: at 1220 a.m. Pacific Island Time, the Comsat system received a signal from the Galaxy Inmarsat-C/GPS system indicating a small deviation from the ship's course to Tarawa, and immediately issued an electronic order to the helm to steer a three-degree starboard correction. The Autohelm dutifully responded, and the computer log shows the steering change.

At 0102, a signal was received from the traffic service in Boston that GPS signals from ships in the vicinity, including yours, indicated a possible future conflict, some six-and-a-half hours away, with an eastbound freighter. A reminder was logged for the computer to monitor radar images for the vessel at the appropriate time. The information was recorded in the ship's log.

At 0235, a fax was received from New York, from the editor at *Adrift* magazine, asking about the whereabouts of an article on the Line Islands for next month's issue. You check the outgoing telecommunications log and note with relief and satisfaction that the entire sixteen-page manuscript of deathless, seagoing prose was sent at 0300.

At 0512 Coordinated Time, not quite an hour ago, the computer received the input signal confirming that the radar had picked up the freighter at sixty-seven nautical miles off the port bow, and that course, speed, and distance calculations indicated that the two vessels would pass more than five miles abeam one another at approximately 0630. Since it's nearly that now, you leap up the companionway with your binoculars, eager for the sight of another ship to remind you that you're not alone on this vast expanse of oceanscape. Sure enough, there she steams, barely visible through the blue-white salt haze of the early morning.

The rest of the night watch was uneventful, according to the computer log, the only other task being to activate the coffeemaker at 0615. The system scrolled through a reminder to check oil and fluid levels in the engine and cooling system, as well as to check bilge pumps and the water-storage level. The system did everything but make breakfast, a task still left to the sailor.

Now, you ask, is all this possible? Or is it just a piece of nautical science fiction?

Today it is indeed possible, on a boat of any size and usually within the budget of the modern yachtsman. All you need is a relatively inexpensive laptop or conventional computer ($1,500–$2,500) coupled with a not-exactly inexpensive ($6,000–$15,000) system that integrates data, voice, and facsimile communications and GPS navigation and can be tied into the Autohelm or other autopilot device, radar, and a few necessary mechanical linkages (Figure 8-1a, 8-1b, 8-1c, 8-1d).

Figure 8-1 (a through d continued on the following pages). These Autohelm components, tied to the seagoing computer, can create a synergistic aid to navigation and boat handling by automating steering, GPS navigation, and charting. (Photos courtesy of Autohelm)

108 The Seagoing Computer

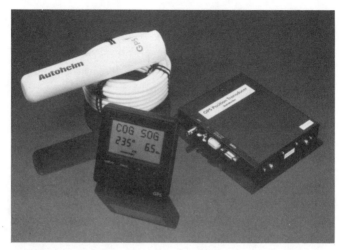

Figure 8-1b. GPS Components.

Figure 8-1c. Autopilot.

Computerizing the Helm and Navigation Tasks   109

Figure 8-1d. Compass.

The Global Positioning System (GPS)

What is GPS?

The Global Positioning System, or GPS, consists of a group of American satellites that orbit the earth twice a day, transmitting precise time and position (latitude, longitude, and altitude) information over discrete frequencies (Figure 8-2). Using a GPS receiver, the mariner can determine his or her location anywhere on the surface of the earth. So capable are these systems for correlating and analyzing the satellite signals that the ocean navigator can determine a fix within ten meters or less.

When fully operational in 1993–94, the complete Global Positioning System will consist of twenty-one

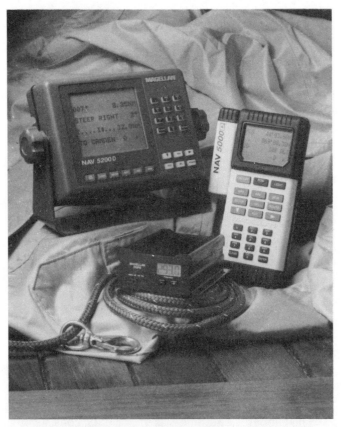

Figure 8-2. Global Positioning Systems, such as these cockpit-mounted (left) and hand-held (right) units, represent the latest in accurate, inexpensive, and sophisticated electronic-navigation systems. They can be connected to the onboard computer to provide even more information and utility. (Photo courtesy Magellan Systems Corporation)

satellites and three spares orbiting approximately 11,000 miles above the earth. These satellites will provide 24-hour-a-day coverage for both two- and three-dimensional positioning anywhere on the earth's surface.

Developed over a period of fifteen years, the $10-billion GPS system is being deployed by the U.S. Department of Defense to provide continuous, worldwide positioning and navigation data to U.S. military forces around the globe (Figure 8-3). The system saw much use during Operation Desert Shield and Operation Desert Storm, where United Nations Coalition forces used hand-held Magellan GPS receivers to determine their position in the vast, featureless desert expanses of Saudi Arabia, Kuwait, and Iraq.

However, its designers and builders realized that GPS has the potential for an even broader commercial application. To achieve this goal, the GPS satellite transmitters send data on two frequencies, one for civilian access and the second encrypted for exclusive military use. The GPS signals are available to an unlimited number of users simultaneously.

GPS technology basically involves correlating very precise time and position information between two or more satellites to arrive at a precise position on the earth's surface. Using location data and atomic clocks that boast an accuracy to within one second every 300,000 years, each satellite continuously broadcasts the exact time, along with its geographic location above the planet. A GPS receiver intercepts these signals and, comparing the precise location of three or more satellites at once, determines the user's location on the earth's surface.

The position information in a GPS receiver may be displayed as longitude and latitude coordinates, Universal Transverse Mercator coordinates, or Military Grid coordinates. To obtain two-dimensional (latitude and longitude) fixes, data must be received from three satel-

112 The Seagoing Computer

lites. Four satellites are required for three-dimensional (latitude, longitude, and altitude) positioning.

Each GPS satellite continuously broadcasts two signals—a commercial Standard Positioning Service (SPS)

Figure 8-3. A Magellan GPS receiver like this was used in a joint U.S.-Soviet-Canadian expedition to locate and film the wreckage of the Titanic. *(Photo courtesy of Magellan Systems Corporation)*

signal for worldwide civilian use, and a Precise Positioning Service (PPS) signal for U.S. military use. The SPS signal will provide a civilian user an accuracy of up to 25 meters (27.5 yards). Because they are so accurate, civilian GPS receivers using the SPS signal are sometimes subjected to Selective Availability (SA) interference by the U.S. government to maintain the optimum military effectiveness of the system. When engaged, SA inserts random errors in the data transmitted by the satellites. As a result, GPS signal accuracy can be reduced to 100 meters, or 110 yards (another thing for which we can all thank the government.)

Industry has addressed this anomaly, however, with a technique called Differential GPS (DGPS), which enables the user to overcome the effect of the government's generous Selective Availability interference and increase the overall accuracy of the GPS receiver. DGPS works like this: One GPS receiver unit is placed at a known location, and the position information from that receiver is used to analyze the SA-induced errors and calculate the appropriate corrections in the position data transmitted by the satellites. This corrected information is then transmitted to other GPS receivers, with a resulting real-time accuracy of 10 to 15 meters. According to Magellan Systems Corporation, a major manufacturer of marine GPS equipment, accuracy as close as three meters can be obtained by using DGPS and post-processing calculations in static positioning.

What this means is that you, standing on the bridge of your yacht, can turn on your hand-held GPS receiver and, without ever picking up a sextant or squinting at a sight-reduction table or putting plotter or pencil to chart, determine your exact position on the earth's surface to within 10 yards! This is true whether you are situated at Pier 39 in San Francisco, or 1,400 miles southwest of Honolulu in the mid-Pacific. It is probably the best blend of computer and radio-navigation technology yet produced.

Galaxy Inmarsat-C/GPS and Comsat C-Link

Trimble Navigation's Galaxy Inmarsat-C/GPS is a new, integrated communications and navigation system designed to take advantage of the Global Inmarsat Consortium, an international network of global satellite-communications services to which twenty-five nations belong.

The Comsat Mobile Communications is introducing a matrix of services that provide voice and discrete data communications using the Inmarsat satellite network.

Though these two companies are by no means alone in developing services based on newly available satellite technology, they are in many respects the only providers of these services. Remember, this is all brand new, especially where smaller and noncommercial vessels are concerned. Therefore, for purposes of this book, Trimble and Comsat have provided descriptions of the services and equipment they offer the marine industry as examples of what is available in seagoing computer technology at the dawn of the truly global telecommunications age.

Trimble Navigation Galaxy Inmarsat-C/GPS

The Galaxy Inmarsat-C/GPS is a compact, integrated system that uses a laptop or conventional computer and a transceiver hookup to process signals from two kinds of satellites—GPS and the global grouping of geosynchronous satellites known as Inmarsat—to provide marine or land-based receivers with exact position data, two-way messaging, maritime safety information, and emergency notification. It was designed not only for larger merchant ships and other commercial vessels, but also with an eye toward the space and budget limitations of the seagoing yacht.

The advantages of such a system are manifold, ranging from the ability of a cruising yacht on extended passages to keep shore parties apprised of its position, a sort of enhanced-safety improvement over the old boatplan, to up-to-the-minute weather, two-way voice and data communications, fax, and weather analysis. A complete system, consisting of an Inmarsat-C transceiver, a GPS receiver, dual-function antenna for both GPS and satellite communication, laptop computer, printer, accessory kit, antenna cable, system power supply, and the proprietary Trimble Navigation integrated computer software, sells for a retail price of $14,280. The Galaxy transceiver alone sells at retail for $7,800. Both prices are FOB Sunnyvale, California.

As with all computer equipment, components, and accessories, prices will undoubtedly fall once the systems have been on the market for a time and the cost of research, development, and marketing has been amortized.

Comsat C-Link

Operating twenty-four hours a day over all oceans, Comsat operators can handle voice, data, fax, telex, electronic mail, and distress calls in 140 languages from any vessel equipped with a laptop computer and the appropriate transceiver and antenna equipment.

The Comsat C-Link system uses satellites in geosynchronous orbit 22,000 miles above the earth's surface that relay data from earth stations located in Southbury, Connecticut, Santa Paula, California, and Anatolia, Turkey. The "ship segment" of the system utilizes the satellite communications terminals, also known as "ship earth stations," or SESs. There are four types of SESs, namely, Inmarsat A, B, C, and M; however, our only concern here is Inmarsat-C, which incorporates the latest SES technology and is designed for vessels under 80 feet. While there has been a wide disparity between services provided via Inmarsat A and B, for larger vessels, and Inmarsat C and M, for smaller boats, Comsat C-Link is now making virtually every service available to small-vessel operators, as well as to operators of large commercial passenger and other marine vessels.

Comsat C-Link provides ship-to-shore, ship-to-ship, and shore-to-ship telephone and data-communications service, and even television transmissions. SeaMail, the company's electronic-mail service, dedicates a full-service system to maritime users. With modem rates of up to 9600 bps, you can transfer data, send and receive e-mail, download (or upload) files to the Bulletin Board System, and receive a variety of news and information services, including the Maripress News Service, containing a digest of top U.S. and international news, sports, and financial information; Marisports, sports news; and radio and television broadcasts.

Comsat will track your progress across any ocean, adding a not inconsiderable level of comfort and confidence to those long bluewater passages. Obviously, such service is not for the strictly traditional sailor, who would probably scoff heartily at such a direct link to shore. But for those wishing the latest and best in communications capabilities in order to work and live aboard while cruising, today's computer/satellite technology certainly opens up a whole new world of possibilities.

"Iron Mike," the Automated Helmsman

What the autopilot is to the airplane pilot can be even more so to the yachtsman, who otherwise would be forced, like the mariners of old, to stand his seemingly endless watch at the helm, hand upon the tiller, eyes squinting, gimbaled on the horizon, ever mindful of the point of sail and the compass, responding to the slightest change of course. Not so today, when fully automated autopilots can be integrated with the computer for flawless navigation and hours, even days, of unattended steering. The computerized automatic helmsman is a crew member that never gripes, never dozes off, never (or perhaps I should use a qualified "almost never") goes overboard, and never has to be fed. The machine also has the rather impudent trait of being able to navigate far more accurately than the most exacting skipper.

The NMEA 0183 standard refers to an industry-wide conforming interface for marine instruments, computers, and accessories. Using this interface system, electronics manufacturers have designed ways to integrate electronic navigation, radar, and steering functions among differing components. Thus, nearly all Loran and GPS navigation receivers sold today can be linked to computerized navigation programs/systems and autopilots. Navigation chart plotters are viewed on the computer terminal, but also linked with the radar set, superimposing a navigation chart image over the radar plot, and the autopilot, transmitting steering information based on satellite nav data, oceanographic chart data, or both. The plot the navigator establishes is there on the computer but is also available to the helm. So all functions are neatly linked to create an integrated whole.

The problem for the yacht owner is sifting through the myriad articles, reports, advertisements, and claims made by manufacturers and dealers for the capabilities and true compatibility of various computer, navigation, radar, and steering devices.

First, be sure that the NMEA 0183 standard is incorporated into the component's design. Without such interface capability, you may be stuck with an instrument or device that cannot be integrated with the rest of your boat's gear.

Second, stick with well-known national or international manufacturers that have a reputation for backing their claims and products.

Third, demand written guarantees and warranties based upon installation by the manufacturer or the manufacturer's representative, and invoke them at the slightest indication of a product's malfunction. This equipment is expensive, and no one can afford to tie up money and space with gear that doesn't work, or doesn't get along with the other members of your boat's electronic "crew." Be sure to ask the same of computer manufacturers, and sellers of other traditionally nonmarine equipment, who may be new at the nautical game.

Don't hesitate to ask for a sea trial. Precautions taken here will save you time, money, and possible hazardous malfunctions at sea.

The following is a list of Inmarsat-C equipment manufacturers. Other makers of related equipment and various navigation receiver/computer/satellite interfaces are listed in the appendices of this book.

U.S. Manufacturers:
Furuno U.S.A., Inc.
(Felcom 10 Maritime Model)
Att: Larry Griswold
P.O. Box 2343
South San Francisco, CA 94083
Tel: (415) 873-9393
Fax: (415) 872-3403

Frotronics, Inc.
(Safecom CM Maritime Model)
Att: Jack Frost
6142 South Loop East
Houston, TX 77087
Tel: (713) 644-6445
Fax: (713) 644-2134

Koden International
(KSC-90 Toshiba Maritime Model)
Att: George Lariviere
77 Accord Park Dr.
Norwell, MA 02061
Tel: (617) 871-6223
Fax: (617) 871-6226

Mackay Communications, Inc.
(EB Nera Maritime Model)
Att: Thomas Finneran
Raritan Center
300 Columbus Circle
P.O. Box 7819
Edison, NJ 08818-7819
Tel: (908) 225-0909
Fax: (908) 225-2848 or (908) 225-4959

Mobile Telesystems, Inc.
Data Lite
(Land Mobile Model)
Att: Christopher J. Henderson
300 Professional Dr.
Gaithersburg, MD 20879
Tel: (301) 590-8526
Fax: (301) 590-8558

Raytheon Marine Company
(JRC Maritime Model)
Att: Mike Mitchell
46 River Rd.
Hudson, NH 03051
Tel: (603) 881-5200
Fax: (603) 881-4756

R-H Trading
(Division of Radio Holland)
Thrane & Thrane
(Maritime and Land Models)
Att: Mark Richards

8931 Gulf Freeway
Houston, TX 77017
Tel: (713) 872-3464
Fax: (713) 946-1115

SEASAT
(SNEC Maritime Model)
Att: Lai Jones, President
Suite 220
7050 Oakland Mills Rd.
Columbia, MD 21046
Tel:
Baltimore, MD (301) 381-0605
Washington, DC (301) 596-2722
Fax: (301) 381-3611

Scientific Atlanta
(Series 9800 [Thrane & Thrane])
Att: Macy W. Summers
4356 Communications Dr.
Norcross, GA 30093
Tel: (404) 903-6168
Fax: (404) 903-6245

Trimble Navigation
(Galaxy Inmarsat-C/GPS Maritime Model)
Att: Don Green
645 North Mary Ave.
P.O. Box 3642
Sunnyvale, CA 94088-3642
Tel: (408) 481-8000
Fax: (408) 737-6057

Non-U.S. Manufacturers:
EB Nera
(Saturn-C)
Att: Mr. Sigurd Th Helland
EB Nera a.s.
Satcom Marine
P.O. Box 91
N-1361 Billingstadsletta
Oslo, Norway
Tel: +47 2 84 47 00
Fax: +47 2 84 46 21
Telex: 71721 UMEB N

Furuno Electric Company
(FELCOM 10)
9-52 Ashihara/cho
Nishinomiya City, Japan
Tel: +(0798) 65-2111
Fax: +(0798) 65-4200, 66-4622, or 66-4623
Telex: 5644-325

Japan Radio Company Ltd.
JUE-65A
Att: Mr. A. Sumita,
Executive Managing Director
Japan Radio Company Ltd.
17-22 Akasaka 2-Chome
Akasaka Twin Towers
Minato-ku
Tokyo 107, Japan
Tel: +81-333584 8755
Fax: +81 333584 8891
Telex: 2425420 JRCTOK J

Philips Radio Communications Systems
(SAFECOM CM)
Att: Mr. Jens Bjarnoe Thostrup
Jenagade 22
Postbox 1818
DK-2300 Copenhagen S, Denmark
Tel: +45 32 88 2222 or +45 32 88 3749
Fax: +45 31 57 29 30
Telex: 31201 PHIL DK

SNEC
(C-MATE)
Att: Mr. R. Dickens
2 Rue de Caen
BP 7
14740 Bretteville–L'Orgueilleuse
France
Tel: +33 31 80 71 22
Fax: +33 31 80 65 49
Telex: 171813F

Thrane & Thrane A/S
(CAPSAT TT 3020 A/B Series 9800)
Att: Mr. Per Thrane, Managing Director
Thrane & Thrane A/S
Tobaksvejen 23
DK 2860 Soborg, Denmark
Tel: +45 31 56 4111
Fax: +45 31 56 2140
Telex: 19298 THRANE DK

Toshiba Corporation
TM 1250 A
Att: Mr. Shuji Funo, Senior Manager
1-1 Shibaura 1-Chome
Minato-ku
Tokyo 105, Japan
Tel: +81 3 3457 3120
Fax: +81 333456-1699
Telex: 22587 Toshiba J

The above information was provided courtesy of:

Comsat Mobile Communications
Communications Satellite Corporation
950 L'Enfant Plaza, SW
Washington, DC 20024
Tel: (202) 863-6567 or (800) 424-9152
Fax: (202) 488-3814 or 3819
Telex: 197800

Manufacturers of Inmarsat equipment must be approved by the international Inmarsat organization. The manufacturers listed have Inmarsat approval, and the information is current as of January 1993.

9

Using the Seagoing Computer for Navigation

One hears a great deal about computers used for navigation. The obvious question is: How does a computer simplify the mariner's navigation tasks?

The simple answer is that the computer can take all the hard work out of navigation. For one thing, the machine is capable of storing all those cumbersome charts, tables, and graphs, making the use of them much easier. For another, all the calculating and solving, all the plotting and interpolating, is done at the speed of light, with little effort on the part of the navigator.

Every day brings new and better computer programs for marine application, and those described here are but a small sampling of their growing numbers. For an idea of what's emerging all the time in marine-navigation computer software, pick up a copy of *Ocean Navigator*, a bimonthly magazine for serious circumnavigators and professional/commercial marine officers. In it you will find advertisements and articles describing the latest in software developed for navigation and pilotage, boat operations, marine finance, training, and more.

The programs below are all commercial software developed for marketing to the marine industry. However, before deciding what programs to install on your system, don't overlook the many fine shareware and freeware software programs available from the on-line electronic databases. This new and rapidly growing phenomenon in the nautical world may actually be the best all-around source of economical software for all marine and non-marine applications onboard. On-line services are listed in Appendix B.

Let's take a look at some of the popular marine computer-software programs available on the market today.

Navigation Programs

J. Henry Celestial Navigation Program

This is a very popular IBM PC-compatible celestial-navigation software program, available for $99 by mail order and through the marine computer catalog of D.F. Crane Associates in San Diego, California. J. Henry is designed to enable reduction and plotting of celestial sights using an IBM-compatible computer. It also provides the beginning or novice navigator with an introductory course in celestial-navigation theory and execution.

If you follow the entertaining, well-written explana-

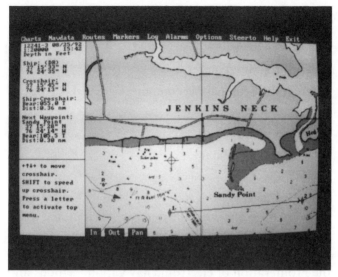

Figure 9-1. A computer-generated electronic navigation chart.

tions and practice exercises using the software and extensive manual, you can't help learning how to use your computer to solve celestial-navigation problems at sea.

The program includes three different sight-reduction methods, along with a perpetual almanac for the sun, moon, 57 navigation stars, Polaris, Venus, Mars, Jupiter, and Saturn. There is also a practice cruise to Bermuda and four hands-on backyard projects. The program incorporates 38 separate tasks for celestial navigation, piloting, coastwise navigation, and current analysis; multiple averaging of sights helps to detect errors. The user can automatically plot the lines of position and fixes on the computer screen, printer, or both.

The J. Henry program comes complete with the necessary printer controls and compass calibrator, as well as a handy reference card and abbreviated reference manual. As a bonus package, the authors include a program called "The World's Easiest Celestial Navigation Course."

The program requires:

- an IBM-compatible computer with diskette or hard drive
- 256 Kb RAM
- DOS 2.1 or later

Traditional Celestial Navigation with J. Henry This tutorial software program is designed to teach traditional celestial navigation using the Nautical Almanac and sight-reduction tables. It is an incontrovertible fact that despite the proliferation of electronic navigation methods and systems, learning to reduce a sight by traditional methods is the best possible backup in case something goes wrong with your computer, and this course is probably the least painful way to learn how to do it.

It provides everything you need, including textbook, interactive software, and instructions from the Nautical Almanac and H.O. 249 for solving simple practice problems. The complete Nautical Almanac and H.O. 249 or H.O. 229 are required if you select the general exercises.

Computer requirements are the same as those for the J. Henry Celestial Navigation Program. Cost is $44.

Macintosh HyperNavigator

This new software is designed for novice or middle-level students of celestial navigation who are also Mac users. It computes GHA, Dec, corrected sextant altitude, intercept, and azimuth for the sun, moon, four planets, and 59 stars. The intercept and azimuth are transferred to an on-screen plotter with three levels of resolution, then a "most probable" position is calculated and a confidence ellipse is plotted around the position.

The Lines of Position (LOP) are computed from published tables and are good through 1995. A perpetual almanac is included for the sun, with almanacs for the moon and stars available to registered owners of the software.

D.F Crane Associates, mail-order catalog dealer for the HyperNavigator program, offers it in two parts, priced at $49 and $79, respectively, plus a $5 license fee. The price includes a special introductory bonus: a bibliography of over 500 articles recently published in leading sailing magazines, and a voyage planner with the dates, distances, and ports for Tania Aebi's circumnavigation as an example.

Requirements:

- an Apple Macintosh computer (any model)

Celestial Navigation Kit

Together with an IBM-compatible PC, this kit from Davis Instruments can help prepare you for a voyage. It can also be used as an educational project, or let you become an armchair navigator. Priced at $129, it includes:

- Mark 3 Sextant. Ideal for training, backup, or beginning navigators, the Mk 3 has been used as the sole means of navigating both the Atlantic and Pacific oceans.
- Artificial Horizon. This lets you practice navigation anywhere, including in your backyard.
- PC Astro Navigator software. This program

provides user-friendly training for dead reckoning, celestial navigation, and navigation plotting. It includes a Nautical Almanac through the year 2024 and a very useful twilight-planning function.

- complete instructions in individual manuals and a "how-to" workbook

Davis PC Astro Navigator Celestial Software

This best-selling $69 program from Davis Instruments is for navigators who know the principles of celestial navigation but would like to be freed from the complexities and drudgery of entering tables and doing the math to reduce and plot sights. Easy to understand and use, it is focused more on the "how-to" parts of celestial navigation than the "why," epitomized by the J. Henry programs.

Like J. Henry, it can be used for both dead reckoning and celestial navigation. PC AstroNavigator can be

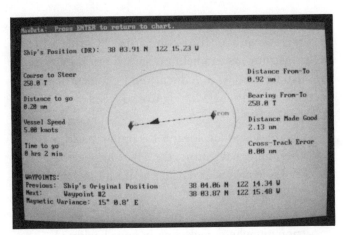

Figure 9-2. A navigation data screen. (Photo courtesy of D.F. Crane Associates, Inc.)

Using the Seagoing Computer for Navigation 131

used as the primary method of performing celestial-navigation sight reductions, or for checking manual calculations while learning traditional methods.

The software can help experienced navigators analyze results from past voyages. The extensive features are menu-driven and user-friendly and include:

- Dead Reckoning(DR). This program keeps track of running fixes and calculates courses and distances made good from previous position and last fixed position. It also maintains a log of all running and fixed-position entries. Its plotting feature allows the most current set of running fixes to be plotted on your computer screen and sent to the printer.

- Celestial Navigation. This calculates noon sights and celestial lines of position (LOP) for any date between 1925 and 2024. Data from the Nautical Almanac is automatically entered, as well as the assumed position from the dead-reckoning log. LOPs, noon sights, and dead-reckoning position can all be plotted to the computer screen or printed; LOPs are automatically advanced to correct position via the use of a running fix log from Dead Reckoning. The program maintains a log of all LOPs and noon sights.

- Nautical Almanac. This provides complete coverage from 1925 to 2024. Just enter date and time and the computer will display declination, Greenwich hour angle (GHA), right ascension (RA), and sidereal-hour angle (SHA; stars only) for Aries, the sun, the moon, Venus, Mars, Jupiter, Saturn, and the 57 navigational stars.

- Twilight Planning. When you enter the date and assumed position, the computer will display the altitude and azimuth of all visible bodies at dawn and dusk.
- Body Finder. Enter the date, time, and assumed position, and the computer will display the altitude and azimuth of all visible bodies.
- Sunrise/Noon/Sunset. Just enter the date and assumed position, and the computer will display sunrise, noon, and sunset.
- Global PC Navigation System. The Global PC Navigation System covers almost every aspect of navigation. It is designed simply, with excellent documentation, and includes a complete celestial-

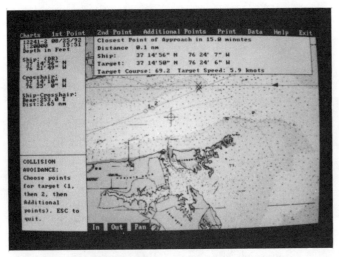

Figure 9-3. The screen display of the ARPA Collision Avoidance computer system. (Photo courtesy of D.F. Crane Associates, Inc.)

navigation training course within the 300-page manual. (H.O. 249 and the Nautical Almanac are required for course exercises.)

The program's functions include: Celestial and Dead Reckoning; Great Circle Calculation; Compass Deviation Control; Twilight Forecast; Star, Moon, and Planet Finders; Night Sky Planispheres; Sun, Star, Moon, and Planet Sight Reductions (Nautical Almanac is in the software); Sight Clustering; and 2-Plot and Running Fixes.

As navigation aids, the program includes Wind and Tide Corrections; Tacking Assistance; Traverse Summation; Beaufort Scale Data; Vertical and Horizontal Sextant Positions; Radio Direction Fix Finder; Compass Error Checks; Temperature, Distance, and Time-Arc Conversions; and Sunrise-Sunset Reporter.

Racing and voyaging are addressed with Racing Prognostications; True Wind From Apparent; Voyage Plans; Passage Logs; Calculation and Plotting Great Circle Courses; Plots and Stores Fixes During Voyages; and Mercator Charts.

For education and training the program offers Star-Finding Tutorial with On-Screen Star Charts and a Traditional Celestial-Navigation Course.

Tide-Prediction Programs

Tide.1 (Rise and Fall) and Tide.2 (Ebb and Flow)

These two programs from Micronautics Corporation enjoy an excellent reputation among oceanographers and commercial operators for accuracy and flexibility. Tide.1 Rise and Fall predicts times and heights of high and low water at all NOAA coastal locations throughout North America and the Caribbean. Tides can be displayed, printed, or plotted, and Calendar Creator Plus can create customized tide calendars.

Tide.2 predicts tidal currents at the NOAA coastal locations in North America. The program displays computed times, speeds and directions of ebb and flood, slack water, and a daily current curve on the screen. The program distinguishes among ordinary reversing tidal stream, rotary currents, and hydraulic currents.

These programs sell for $99 each and provide a full range of output options, including printed results and data tables for other purposes such as plotting, tabulating, and word processing. Calendar Creator Plus, which sells for $69, and can be used to make custom tidal-current calendars. Data is provided for the current year. Annual data updates are available at half the original purchase price.

Requirements include

- an IBM-compatible computer with a diskette drive and monitor
- 384K RAM
- DOS 2.0 or later

A printer is optional. Calendars require Calendar Creator Plus.

The region list includes (one region of your choice with Tide.1 and Tide.2):

Region	Program	Description
R10	Tide.1 or Tide.2	New York or New England
R12	Tide.1 or Tide.2	Mid-Atlantic Coast
R14	Tide.1 or Tide.2	Southeast and Gulf Coast
R16	Tide.1 or Tide.2	West Coast and Hawaii
R18	Tide.1 or Tide.2	Alaska and British Columbia
R20	Tide.1 only	Central America and Caribbean
R22	Tide.1 only	Canada East Coast

Accutide 1.2

Accutide was developed by Dr. Clyde Ford to fill the need for a tidal height and current program that could display and print tide information in easy-to-use graphic formats. Besides being simple to use and offering stunning graphics to illustrate height and ebb-and-flow information, two other design features make Accutide a particularly good value:

1. You can display both tide rise and fall and tide-current information within the same program. Switching between a visual representation of tides and currents is as simple as pressing a key.

2. Accutide data is available for multiple years. Official NOAA sampling-station data was initially provided through 1992, as NOAA has announced it will be making changes affecting the accuracy of 1993 sampling-station data. As of early 1993, a low-cost, updated data subscription covering 1993–96 is available. Data will also be made available for subsequent three-year periods,

and the dealer or reseller will issue NOAA updates periodically, saving the purchaser the expense and inconvenience of ordering new software every year.

Accutide lets you select the NOAA tide or current sampling station you want by region (one region is included; others—or all regions—are available at reduced prices).

Tidal information for the time period you choose can be presented in summary or detail on your computer screen, printed, or stored to a disk file.

For locations in the United States, Accutide automatically determines the start and end of Daylight Savings Time and incorporates this determination into the predicted times of tidal height and tidal-current flow.

Information display includes: Display as a Bar Graph; Turn Slow Water On or Off; Display a Line Graph; Show Average Statistics; Print Screen; Modify Graph Colors; Show Current Data Only; Show Grid Lines Only; or Show Tide Data Only. You can tailor the program to reflect: Initial Current Station; 12- or 24-Hour Clock; Initial Region; NOAA or System-generated numbers; Initial Tide Station; Slow Water Speed; or Sunrise Station.

Accutide sells for $189 for one region, or $429 with all regions included.

The program requires:

- an IBM-compatible computer with a diskette drive
- 512 Kb RAM
- DOS 2.0 or later

The region list includes:

Region	Description
R1	Canada East Coast and New England
R2	Mid-Atlantic Coast
R3	Southeast, Gulf Coast, and Caribbean
R4	West Coast and British Columbia
R5	Alaska and Hawaii

Harbormaster

Harbormaster is a comprehensive tide- and current-prediction program for the Macintosh that includes 1991, 1992, and 1993 data for the East, West, and Gulf coasts of the United States, Alaska, and Hawaii. The program sells for $149.

Users of Harbormaster include barge operators, recreational sailors, the *Los Angeles Times*, and the hazardous-materials group of NOAA, which gives a fairly diverse testimonial to its versatility and acceptability.

Tide locations can be listed alphabetically, by area, or by latitude or longitude; tidal heights can be displayed in feet, inches, meters, or centimeters; and current velocity can be shown in knots or feet per second.

A calendar feature displays data between any two dates over a three-year period. Standard or Daylight Savings Time can be selected and then shown on either a 12- or 24-hour clock display. Annotation is shown for high and low tides, and ebb, flood, and slack currents. The Harbormaster window can be saved in a PICT file, sent to an ImageWriter or LaserWriter, and copied into the clipboard for composition in paint or drawing programs.

138 The Seagoing Computer

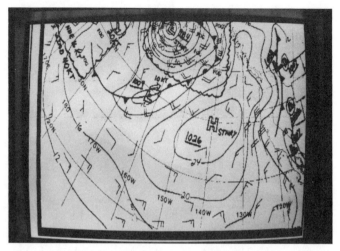

Figure 9-4. A sample of NOAA's Synoptic Surface Analysis Chart as displayed on a computer screen. These, along with most NOAA charts, can now be obtained on media that can be used on either a PC or a notebook computer with a VGA screen display. (Photo courtesy of D.F. Crane Associates, Inc.)

Harbormaster requires:
- a Macintosh Plus, SE, Classic, LC, SE/30, or II series

Charting Programs

Micro Mariner: Electronic Charting Version 2.0

Micro Mariner is a complete program of electronic charting software that allows your IBM-compatible computer to accept input from GPS, Loran, or SatNav, then displays your position on exact electronic reproductions of paper navigation charts. The program sells for $995.

Micro Mariner offers significant advanced features. These include hardware independence, which means that the operator may choose whatever IBM-compatible computer fits needs and budget. Micro Mariner requires:

- only 640Kb RAM
- a hard drive of any size, a 1.44-megabyte diskette drive
- either EGA or VGA graphics

For speed and efficiency, the navigation charts reside on the hard-disk drive, where they can be retrieved, zoomed in and out, and viewed at speeds of ½ to 4 seconds, depending on the speed of the computer. Some charting systems with CD ROM discs or other proprietary storage devices can take 15 seconds or more to perform these basic functions.

Instant help is available. The Operations Manual is actually in the software itself, and to get help at any time, just look in the Information Window that is always on the screen, or press the Help key for a more detailed explanation.

The Micro Mariner program and navigation charts are sent to you on standard 1.44-megabyte computer diskettes. There are no special cartridges or CD ROM discs, and no need for additional peripheral devices. Besides being convenient, the floppy diskettes serve as backups to the chart library on your hard drive.

Real charts are used; the charts are optically scanned, by the developer, from U.S. NOAA, DMA and other countries' hydrographic paper charts. All details on the original paper charts are shown in color, including aids to navigation, depth contour lines, all soundings, and all information printed on the landmasses.

Individual charts are employed. Electronic charts, like paper charts, are ordered by their number. You decide what charts are most important for the area you need, then order only those that you want.

The user has access to periodic updates by subscribing to Local Notice to Mariners and using the notation tools provided with Micro Mariner to make changes right on the electronic charts.

Charts are available for any place on the wetted surface. Whether you contemplate going to Vanuatu, Tonga, Greece, or Norway, if the required electronic chart is not available, the developer will make it especially for you at no additional charge.

Micro Mariner provides three zoom levels per chart. In conjunction with the U.S. government standard of up to 10 different scales of chart coverage for a geographic area, up to 30 levels of zoom can be used for any charted area.

A plotter feature makes it possible to take advantage of the navigation features of Micro Mariner without ordering any electronic charts. A standard function, the plotter feature will show the ship's position on a user-sized blank screen complete with waypoints, markers, alarm zones, and other program features.

In addition to interfacing with GPS, LORAN, and/or SatNav for electronic navigation, Micro Mariner also supports traditional dead-reckoning navigation and position fixes from radar range and bearings, as well as multiple compass bearings.

For users with special requirements, Micro Mariner can be adapted by the developer to the particular needs.

Micro Mariner Surveying/Scientific System

In addition to all the features of the standard Micro Mariner system, the $2,695 Surveying and Scientific System can log data from a variety of different instruments such as depth sounders and magnetometers. Manufacturer-specific digital data output is first converted to standard NMEA data sentences and then logged (at user-specified intervals) to the computer's hard disk for later analysis.

The Surveying and Scientific System is both software and an external interface that accepts up to four different data inputs and multiplexes them into one RS232 serial port on the PC. The system can be modularly expanded to accept up to 30 inputs.

System features include:

- Over 50 on-line help screens
- Operation from keyboard, mouse, or trackball
- Easy-to-understand pull-down chart reproductions
- Optically scanned, exact reproductions of paper charts
- Custom plotter feature
- Up to 30 levels of zoom
- Panning within charts and from one chart to another
- Next chart chosen automatically before vessel leaves the screen
- Continuous display of vessel position and date/time information
- NMEA 0183 interface for Loran, SatNav, and GPS receivers

- Dead reckoning, radar range and bearing, and visual fixes
- Electronic dividers for measuring distance and bearing
- Extensive route planning and voyaging functions
- Steer-to function for navigation between waypoints on a route
- Nautical miles, statute miles, or kilometers
- True or magnetic bearings
- Automatic worldwide magnetic variation
- Automatic computation of bearing, distance, speed, and ETA information
- Manual and automatic ship's log recording, display, and printing
- Automatic log interval setting by distance and time
- Logged data can be exported to user programs
- Unlimited number of waypoints and routes
- Eight differently shaped and colored markers
- Names and/or numbers can be given to waypoints, routes, and markers
- Printer capability for charts, ship's log, and routes
- Navigational Date Display showing:

Date and time	Lat/lon of the vessel
Distance made good	Vessel speed
Course to steer	Distance to go
Cross-track error	Time to go

  Lat/lon and name of previous bearing/distance from/to waypoints and next waypoint

The system requires:
- an IBM-compatible computer with one hard drive

Using the Seagoing Computer for Navigation 143

and one 1.44-Mb floppy drive
- at least 640 Kb RAM
- DOS 2.1 or greater
- one RS232 serial port
- EGA or VGA graphics
- Loran, SatNav, or GPS

Navlink Version 1.2: Computer to GPS/LORAN/SatNav Interface

This new $99 hardware and software product brings navigation from any GPS, LORAN, or SatNav into an IBM-compatible PC. The computer screen then acts as a repeater for the ship's navigation receiver(s) to show pertinent navigation data, as well

Figure 9-5. A gray-scale satellite photo image, typical of the display quality available on the onboard PC or notebook with VGA display. (Photo courtesy of D.F. Crane Associates, Inc.)

as a plot of the vessel's track in real time.

Navlink's hardware is a self-contained NMEA 0183–to–RS232 interface that connects a navigation receiver to the serial port of the computer. Clearly written and illustrated instructions make the two-wire connection east to install.

The software provided with Navlink is easy to use, yet offers some powerful features that can help turn any computer into a very sophisticated navigation instrument.

For example, you can first plan a route either by entering waypoints in plain English or by using Navlink's extensive library of Coast Guard and DMA light-list waypoints. You can then watch your vessel's track move across the screen as you steer along the route. The display shows present latitude/longitude, speed over the ground, name of the start and destination waypoints, and distance and bearing between position and destination, as well as start and destination waypoints, cross-track error, time to go, and estimated time of arrival.

A logging feature can record your trip, which you can later replay showing plot, distance traveled, maximum speed, and start/stop and elapsed time of the voyage.

As a bonus, useful information such as VHF and SSB channels, weather fax frequencies, a multi-time-zone clock, and a calculator are also included in the software.

The program requires:

- an IBM-compatible computer with a diskette drive
- 256K RAM
- DOS 3.3 or later
- CGA, EGA, or VGA graphics

- and Loran, SatNav, or GPS with NMEA 0183 output

Waypoint Management Software

If you have a collection of waypoints you need to refer to, such as the locations of fishing spots, diving coordinates, or other charted places of interest, this versatile $69 software program may be just what you need.

It can help you organize and manage your list of waypoints, and even create maps from the information. With 640K RAM, the program will handle files with up to 7,000 readings.

You first enter the position of your waypoints in lat/lon, Loran C, or even old Loran A TDs into the program, along with a description and four-position rating code that you make up to reflect the characteristics of your waypoints.

To make data entry easier and faster, the program lets you "tailor" soft-key definitions of words or phrases that you often use to describe your waypoints. The program will then let you quickly search and sort by lat/lon, TDs range and bearing, or rating code to find what you are looking for among a large amount of information. The results can be in the form of a printed list or a graphic display of selected waypoints on a lat/lon grid.

This software can also help you produce trip plans that include the range, bearing, and distance from each waypoint to the next, and the total distance for the trip.

In addition, charts of your waypoints on lat/lon grids can be produced. Symbols can be used to give special significance to locations on the chart, and islands and coastline can be drawn right on the screen using lat/lon coordinates.

Special routines allow for converting Loran A and Loran C to lat/lon, and for importing and exporting ASCII files.

The program requires:

- an IBM-compatible computer with diskette drive
- 384K RAM
- DOS 2.0 or later
- CGA, HGA, EGAS, or VGA graphics; a math co-processor is optional

Navigate!

Navigate! is a comprehensive, digital chart-navigation system designed for sailing yachts, power yachts, and commercial mariners.

The software sells for $395 for the monochrome-display version and $695 for the color version, and it runs on all Macintosh monochrome or color computers, including Mac laptops. If you already own a Macintosh and a navigation receiver with NMEA 0183 output, your only additional investment to get the benefits of electronic charting is the Navigate! software.

Charts are hand-digitized from NOAA paper charts and distributed on Macintosh 3½-inch diskettes. The chart data is organized much as on NOAA paper charts, with each chart diskette containing as much shoreline information as seven or eight paper charts, depending on geographic area. All water- and land-navigation aids are included.

The user can zoom a chart through four levels, from close-in harbor to offshore detail. An interface provided with the program ties in a Loran, SatNav, or GPS receiver to provide an automatic display of the vessel's position range and bearing to destination, ETA, course made good, and cross-track error.

Navigate! specifications include digital chart data showing vector contour with nav aids and vessel position at, typically, 1:160,000, 1:80,000, 1:40,000, and 1:20,000; offshore charts are available at lower scales. Scale-to-scale zooming is provided at four levels and each chart data disk contains shoreline information comparable to a typical NOAA NOS geographic area.

The chart presentation provides land contour and nav aids including: Lat/Lon Grid; Loran Data; Pointer Position (Lat/Lon); Vessel Position (Lat/Lon); Destination Waypoint; Numbered Log Roster; Magnetic Variation; Graphic Overlay Filter; Range to Destination; Planned Route; Bearing to Destination; Reverse or Reciprocal Route; Course Made Good; Logging Interval; Cross-Track Error; Magnetic Variation Control; Automatic Logging of Position; Automatic Range/Bearing; Speed, Time, Display of Vessel; Calculation Between Waypoints; and Automatic Calculation of ETA at Any Waypoint.

Nav aids are divided into fixed aids, floating aids, and dangers, and categories can be selected individually. When any aid is selected, an on-screen information window displays latitude/longitudes and characteristics: light, bell, gong, and flash sequence.

Real-time tracking of vessel position and performance is possible with the use of a special interface cable provided with Navigate! The feature requires a RS232 serial port on the computer and an NMEA 0183 data port on the Loran, SatNav, or GPS.

Version 3.0 of the program includes a universal (worldwide) plotting sheet, a digital photo function, floating windows, import/export of waypoint and logpoints, the ability to plan multiple routes simultaneously, user preferences for waypoint/logpoint color, a log annunciator, and selectable range and bearing labels.

The Navigate! program requires:

- a Mac, MacPlus, Mac SE series, or Mac II with one 800K diskette drive; a hard drive is optional
- 512K RAM

This represents a fairly broad sampling of the computer programs available for PC and Macintosh users for celestial-navigation functions, charting, sight reduction, tide and current forecasts, and other marine-specific applications. Installed on a computer already equipped with all of the data-processing and business functions discussed earlier, they transform the onboard computer into a fully integrated working system for enhancing productivity, efficiency, and safety at sea.

Summary

Through the magic of computer technology, the capability to create a productive, profitable, and ideal life aboard is now well within your reach. Through the relatively simple and inexpensive act of installing a computer aboard your boat, you have taken a major step toward becoming a member of that small but growing society of people who have begun to merge the nautical community with the electronic global village, and who now may live and work more effectively at sea.

If, as scientists now say, man's future lies with the oceans of the world, that future will best be enjoyed by those situated on the leading edge of technology, and who are farsighted enough to put it to work for them.

GPS Equipment Manufacturers

The following GPS manufacturers and dealer/distributors handle equipment designed for integration into a computerized marine-navigation system.

B&G
3118 S. Andrews Ave.
Fort Lauderdale, FL 33316
(305) 463-7171

FURUNO USA
Box 2343
S. San Francisco, CA 94083
(415) 873-9393

INTERPHASE
1201 Shaffer Rd.
Santa Cruz, CA 92060
(408) 426-2007

MAGELLAN
260 East Hunter Dr.
Monrovia, CA 91016
(818) 358-2363

MAGNAVOX
2829 Maricopa St.
Torrance, CA 90503
(213) 618-1200

MICROLOGIC
9610 DeSoto Ave.
Chatsworth, CA 91311
(818) 998-1216

NAVSTAR
1500 N. Washington Blvd.
Sarasota, FL 34236
(800) 486-6338

NORTHSTAR
30 Sudbury Rd.
Acton, MA 01720
(508) 897-6600

PRONAV
1120d Thompson Ave.
Lenexa, KS 66219
(913) 599-1515

RAYTHEON MARINE
46 River Rd.
Hudson, NH 03051
(603) 881-5200

ROBERTSON-SHIPMATE
400 Oser Ave.
Hauppauge, NY 11788
(800) 645-3738

SI-TEX
P.O. Box 6700
Clearwater, FL 34618

TECOM INDUSTRIES
9324 Topanga Canyon
 Blvd.
Chatsworth, CA 91311
(818) 341-4010

TRIMBLE NAVIGATION
645 North Mary Ave.
Sunnyvale, CA 94088
(800) 874-6253

B

Telecommunications and On-line Services

The following interactive electronic information services provide information, reference, electronic-mail, financial, news, entertainment, and marine-oriented Special Interest Groups; some provide Internet access, and other services to the computer user. To the extent possible, the listings include information on the nature of these services and their published means of access.

The list is incomplete, since new services spring to life every day and because an effort was made to include only those services that offer the most to the marine computer user.

At least two of these on-line services, CompuServe Information Service and Delphi, have special areas specifically for yacht owners and other marine interests. The Sail BBS is oriented exclusively for marine interests.

AMERICA ON-LINE
1-800-227-6364

America On-Line offers a full array of services including e-mail, Internet access and the usual message forums, Special Interest Groups (SIGs), news, weather, travel, and shopping services. AOL has a large games area as well as a vast database file area. It is, at this writing, the only on-line service that offers a free software starter kit, and the only one that supports Windows screen graphics, though this will soon become the standard for all such services.

AOL is marketed to IBM PC clone and Macintosh users free of initial sign-up or initiation fees, though the fee structure for the two types differs oddly. IBM PCs and clones are billed at $7.95 per month, with two free hours of connect time and additional time billed at $0.10 per minute. Mac users pay only $5.95 per month but get only one free connect hour, after which they pay $5 per connect hour. A customer service representative would say only that the fee structure was part of a marketing strategy.

It's fair to say that almost all of the on-line services employ these strange marketing tactics as they vie for supremacy in the hotly competitive marketplace. Eventually, as the interconnected computer becomes another necessary utility, these vagaries will disappear and the cost will be standardized, like long-distance telephone or electric-utility rates.

BIX
(800) 225-4229
(617) 491-5410 for 2400-bps access; 9600-bps access can be obtained via SprintNet

To log on, set modem to 7 bits, even parity, and 1 stop bit. When connected, enter BIX at the "login:" prompt. When the "Name" prompt appears, type "Bix.Byte" and the menu system will take you from there.

Bix originally began as an electronic on-line service for programmers, expert computer users, and other techies who subscribed to *Byte*, a monthly computer magazine. BIX is now owned and operated by General Videotext Corporation, the same Boston organization that owns and operates Delphi.

Bix provides abstracts and full-text content of articles appearing in *Byte* magazine, e-mail, a variety of interest-group areas, full Internet access, and a gateway to other services through Internet.

The Bix introductory surcharge is $1.20 per hour between 6 p.m. and 7 a.m., and $6 per hour at other times.

To connect with Bix at 9600 bps via SprintNet, dial your local SprintNet 9600-bps number and enter C 834 10100 at the "@" prompt.

For more info, type JOIN ASK.BIX/SPRINTNET while you're on Bix.

COMPUSERVE INFORMATION SERVICE (CIS)

CompuServe Information Service representatives can be reached by dialing the company's Customer Service Center at 1-800-848-8990.

While underway, you can access CompuServe Africa via CSIR-Net (a packet-switching network) at a cost of approximately $18 per hour, U.S. (Note that billing is in local currency.)

CIS subscribers of the nautical persuasion may access the Sailing Forum by typing GO SAILING at any prompt.

For more log-on information, type GO LOGON or call (012) 841-2530 in South Africa and 011-27-12-841-2530 everywhere else.

DELPHI

Operated by General Videotext Corporation, Delphi is a full-service on-line information service that also has a full-service yacht/marine Special Interest Group (SIG) called the Delphi Yacht Club.

To access the Delphi Yacht Club, instruct your computer/modem to dial 1-800-365-4636. At the "Username:" prompt, type JOINDELPHI and press the Return key. At the "Password:" prompt, type YACHTCLUB and press the Return key. You will then be guided by on-screen prompts to supply the usual information, i.e., name, address, etc., and will be issued a permanent password.

Once you're on line, simply type "Yacht" at the Entertainment and Games menu and you will be escorted into the Delphi Yacht Club, with its Message Forum, Database File

Area, Conference Area, Workspace, and Electronic Mail system.

Delphi can be accessed for as little as $1 per hour, exclusive of telephone connect charges, such as SprintNet and Tymnet, through the company's 20/20 Plan. This enables the subscriber to obtain up to twenty hours of non-prime on-line time, between 7 p.m. and 7 a.m., for $20.

DOW JONES NEWS RETRIEVAL
1-800-522-3567, ext. 279 (U.S.)
1-609-520-8349, ext. 279 (New Jersey and overseas)

As the name certainly suggests, Dow Jones News Retrieval is aimed primarily at business interests. But as such, it may be a particularly valuable choice for those intending to live and work onboard. This is particularly true if the business conducted aboard involves the financial markets, or if extensive management of personal or business investments must be done afloat.

Dow Jones News Retrieval charges $29.95 for the initial start-up of the service. A rather complicated fee structure then ensues, ranging from $1.20 per minute during "prime time," or business hours, to $0.15 per minute, plus $0.30 per 1K of information received, after 6 p.m.

The company advertises approximately 60 business and financial services, including:

Breaking Business News

Updates on Competition

Investment Analysts' Reports

Detailed Corporate Profiles

Financial Overviews

P/E Ratios

Current Stock and Bond Quotes

Historical Stock Quotes

Company VCR. Industry Performance

Instant "Background" Information

Income Statements

Balance Sheets

SEC 10K and 10Q Data

SEC Insider-Trading Activity
Earnings Reports and Forecasts
An On-line Business Library

Dow Jones also makes available on line the full text of its *Wall Street Journal* daily business newspaper, as well as the full text of articles from *Barron*'s, *Business Week*, *Forbe*s, and *Fortune* magazines and the daily *Washington Post*. All in all, this highly specific and rather expensive service could be of great value to the yacht owner living aboard and managing a financial portfolio or business.

GENIE
1-800-638-8369 (U.S.)
1-800-387-8330 (Canada)
Set modem to 300, 1200, or 2400 baud, half duplex, 8 data bits, 10 stop bit, and no parity. At the "U#" prompt, enter XTX99463,GENIE and then press Return. You will be guided by menus and prompts through the initial sign-up process.

GEnie, an on-line service of General Electric Information Services Corp., is similar in size and content to CompuServe, Delphi, America On-line, and Prodigy. It offers a wide range of information and services, among them e-mail (called GE Mail) and other messaging services, and news, including NewsGrid, a matrix of news stories from various news wire services around the world; NewsBytes News Network, a news roundup; Dow Jones News Retrieval; and others.

The service also carries *Grolier's Encyclopedia*; EAASY SABRE, an on-line air-travel reservations system that is an adjunct of American Airlines' SABRE reservations system; and various financial brokerage services. There are forum areas for a host of interests, including all the major computer hardware and software manufacturers, and even a medical-research and information area.

Cost for the service ranges from a flat rate of $4.95 per month for the company's GEnie*Basic service during non-prime time, from 6 p.m. to 8 a.m. local time, to $6 per hour non-prime time and $18 per hour for prime time for its Value Services. The value services are aimed primarily at the business-oriented user and offer such ad valorem features as Dow Jones News Retrieval, OAG, QuickNews, and others.

OFFICIAL AIRLINE GUIDES (OAG), Electronic Edition
1-800-323-3537, ext. 024

The Official Airline Guides Electronic Edition enables the computer user to look up airline schedules, routes, fares, and discounts, just as a travel agent would do. The service also includes hotel, automobile-rental, and other travel-related information, for use in planning business and pleasure travel.

PRODIGY
1-800-776-3449

Prodigy, an on-line service of the Sears network, offers many of the typical information, news, and messaging services. But it also bombards the user with on-screen advertising screens and banners from which there is no relief. Although the service has its positive features, including financial, messaging, and other info services, it is, frankly, of questionable value for the serious marine computer user, whose potential needs would quickly outstrip Prodigy's offerings.

THE SAIL BBS

The Sail BBS (Bulletin-Board System), located in Maryland, is one of the small number of emerging electronic on-line services aimed specifically at the marine interest groups. The Sail BBS, as distinct from the commercial on-line services described above, has no hourly connect fees; the only charge is for the long-distance call.

The Sail BBS can be reached at (410) 643-1466. Set your computer's modem up to dial the number and log on at 8 data bits, 1 stop bit, and no parity. Select access at 300, 1200, 2400, or 9600 bauds. Once your modem has connected, you will be guided through your on-line session by a simple, easy-to-use set of menus.

The Sail BBS offers, in addition to a vast nautical database-file area, news and feature articles on sailing, power yachting, racing, and other areas of interest; general news articles from *USA Today*; classified ads in a number of marine areas; e-mail and message areas; yacht club news and announcements; and a number of other services.

The Sail BBS is new, and you can expect it to grow rapidly as more mariners become aware of it, and as more and more marine enthusiasts become equipped with computers and modems.

C

Manufacturers and Dealers of Marine-Computer Hardware and Software Systems

The following computer hardware and software manufacturers, distributors, and dealers offer computer-related products to the marine industry. Where there is a specific brand name or brief description of the product available, it is given at the end of the individual listing. The list contains overlap, since a number of marine manufacturing companies are now involved in more than one aspect of the burgeoning industry of computerizing marine systems, activities, and tasks.

Programmable Calculators, Computers, and Systems Hardware

AUSCO
15 Goose Pond Rd.
Tabernacle, NJ 08088
(609) 268-3081
 Computer interface for nav/instrument data and LCD displays.

D.F. CRANE ASSOCIATES
710 13th St., No. 209
San Diego, CA 92101
Tel: (619) 233-0223
Fax: (619) 233-1280
 "Sea PC" marine personal computer; weatherfax and satellite-photo interfaces.

MARISYS
P.O. Box 810414
Boca Raton, FL 33481
Tel: (407) 272-3490
Fax: (407) 272-3485
 Waterproof PC and Macintosh system and software.

OXKAM INSTRUMENTS
26 Higgins Dr.
Milford, CT 06460
Tel: (203) 877-7453
Fax: (203) 878-0572
 MS-DOS PC computers and interface.

Navigational Weather, Tide Prediction, and Other Software

ACCU-WEATHER
619 W. College Ave.
State College, PA 16801
Tel: (814) 234-9601
 "Accu-Data" PC-accessible weather database.

ANDREN SOFTWARE
P.O. Box 33117
Indialantic, FL 32903-0117
Tel: (407) 725-4115
 The "Loran Program" for IBM PCs.

Manufacturers and Dealers 159

AQUA LOGIC, INC.
c/o Pro Sports Marketing
60 Woods Edge Ct.
Warrenton, VA 22186
Tel: (703)347-1665
"Tidal Logic" tides-prediction program.

AUSCO
15 Goose Pond Rd.
Tabernacle, NJ 08088
Tel: (609) 268-3081
Racing programs.

AUTOMATION PLUS
MicroMarine
P.O. Box 11837
Ft. Lauderdale, FL 33339
Tel: (305) 351-9488
"Passage Maker" cruise-planning software.

BASIC MARINE
P.O. Box 3536
Annapolis, MD 21403
Tel: (410) 268-4619
Integrated navigation and yacht-management software programs.

CELESTAIRE
416 S. Pershing
Wichita, KS 67218
Tel: (800) 727-9785
Fax: (316) 686-8926
"PC Navigator" for IBM, "HyperNavigator" for Macintosh, as well as many other items of hardware and software and marine publications.

COMPUSAIL
22120 Parthenia St.
Canoga Park, CA 91304
Tel: (818) 340-8851
Navigational and performance software programs.

D.F. CRANE ASSOCIATES
710 13th St., No. 209
San Diego, CA 91304
Tel: (619) 233-0223
Fax: (619) 233-1280
"Micro Mariner" chart, weatherfax, and satellite-photo program.

DAVIS INSTRUMENTS
3465 Diablo Ave.
Hayward, CA 94545
Tel: (510) 732-9229
Fax: (510) 732-9188
"PC Astro Navigator" celestial-navigation and sight-reduction software program.

FAIR TIDE TECHNOLOGIES
18 Ray Ave.
Burlington, MA 01803
Tel: (617) 229-6409
Fax: (617) 229-2387
"Navigate!" charting for Apple Macintosh and IBM.

INFOCENTER
P.O. Box 47175
Forestville, MD 20747
Tel: (301) 420-2468
Celestial-navigation programs for Sharp hand-

helds, "CN" nautical almanac for PCs.

MAPTECH
2225 Sperry Ave.
Suite 1000
Ventura, CA 93003
Tel: (805) 654-8006
   CD-ROM charts and "TruChart" plotting with radar overlay.

MICRONAUTICS
P.O. Box 1017
Rockport, ME 04856
Tel: (207) 236-0610
   "Tide 1: Rise and Fall" tide-prediction program, "Tide 2: Ebb and Flow" current predictions for IBM PCs.

NAUTASOFT
Box 282
Rockland, DE 19732
Tel: (800) 999-5221
   "MacTides" tidal prediction for Macintosh.

OCEANSOFT
P.O. Box 1598
Hillsboro, OR 97123-1598
Tel: (503) 693-0747
Fax: (503) 693-1822
   "WayMaker" cruise-planning software program for MS-DOS computers.

OCKAM INSTRUMENTS
26 Higgins Dr.
Milford, CT 06460
Tel: (203) 877-7453
Fax: (203) 878-0572
   "Ockam Soft" performance, racecourse, and tactical software programs for MS-DOS computers.

SOFTWARE SYSTEMS CONSULTING
615 S. El Camino Real
San Clemente, CA 92672
Tel: (714) 498-5784
Fax: (714) 498-0568
   "PC HF Facsimile" SSB reception, "PC SWL" digital communications, and "PC Goes/Wefax" satellite images.

ZEPHYR SERVICES
1900 Murray Ave.
Pittsburgh, PA 15217
Tel: (412) 422-6600
   "Tidemaster" for PCs.

Hand-held Navigation Calculators and Computers

AP SYSTEMS
7461 Pollock Dr.
Las Vegas, NV 89123
Tel: (702) 361-7676
 "Navcom"/"HP-41CX" celestial and piloting computer, "HP-48SX" navigation.

BASIC MARINE
P.O. Box 3536
Annapolis, MD 21403
Tel: (410) 268-4619

CALCULATED INDUSTRIES
22720 Savi Ranch Pkwy.
Yorba Linda, CA 92687
Tel: (800) 854-8075
Fax: (714) 921-2799
 "Measure Master II" calculator.

CELESTICOMP, INC.
8903 SW Bayview Dr.
Vashon, WA 98070
Tel: (206) 463-9626
 Celestial-navigation computer.

CONNEX ELECTRO SYSTEMS
P.O. Box 1342
1602 Carolina St.
Bellingham, WA 98227
Tel: (206) 734-4323
Fax: (206) 676-4822
 "Tide Finder" tide and current software program.

INFOCENTER
P.O. Box 47175
Forestville, MD 20747
Tel: (301) 420-2468
 Sharp hand-held computers with celestial and piloting software programs.

MERLIN NAVIGATION
1520 22nd Ave. E.
Seattle, WA 98112
Tel: (206) 329-8574
Fax: (206) 329-8574
 "Merlin II" celestial-navigation computer.

C. PLATH OF N. AMERICA
(Division of Litton Systems)
222 Severn Ave.
Annapolis, MD 21403
Tel: (410) 263-6700
 "Tamaya NC-99," "Weems & Plath-Galileo."

TAMAYA TECHNICS DISTRIBUTOR
C. Plath N. America
222 Severn Ave.
Annapolis, MD 21403
Tel: (301) 263-6700

WORLDWIDE YACHTING
P.O. Box 3485
Apollo Beach, FL 33570
Tel: (813) 645-8760
   "Quartermaster"/"Psion"
celestial computer.

Electronic Charting and Plotters

APELCO MARINE
ELECTRONICS
(A Raytheon Company)
46 River Rd.
Hudson, NH 03051
Tel: (603) 881-9605
Fax: (603) 881-4756
   "Loran-See" Loran/LCD
chart display.

AUTOHELM
(A Raytheon Company)
46 River Rd.
Hudson, NH 03051
Tel: (603) 881-5838
Fax: (603) 881-4756
   "Navcenter."

BROOKES & GATEHOUSE
P.O. Box 308
New Whitfield St.
Guilford, CT 06437
Tel: (203) 453-4374
Fax: (203) 453-6109
   "B&G," "Network Chart."

C-MAP/USA
P.O. Box 1609
Sandwich, MA 02563
Tel: (800) 424-2627

   Electronic-chart cartridges.

CETREK WAGNER EAST
300 Oak St.
260 Corporate Park
Pembroke, MA 02359
Tel: (617) 826-7497
Fax: (617) 826-2495
   "Chartpilot."

DATAMARINE
INTERNATIONAL
53 Portside Dr.
Pocasset, MA 02559
Tel: (508) 563-7151
   "Dart," "Chartlink."

FURUNO
271 Harbor Way
S. San Francisco, CA 94080
Tel: (415) 873-9393
Telex: 331-419WUI
   Video plotter.

GARMIN
11206 Thompson Ave.
Lenexa, KN 66219
Tel: (913) 599-1515
Fax: (913) 599-2103
   "GPS Map" CRT display

with GPS and "C-Map" navigation charts.

ICOM
2380 116th Ave. N.E.
Bellevue, WA 98004
Tel: (206) 454-8155

MAPTECH
2225 Sperry Ave.
Suite 1000
Ventura, CA 93003
Tel: (805) 654-8006
  "Tru-Chart" software with CD-ROM charts and radar overlay.

MICROLOGIC
9610 DeSoto Ave.
Chatsworth, CA 91311
Tel: (818) 998-1216
Fax: (818) 709-3658
  "MasterChart," Admiral GPS system.

MOTOROLA GEG
8201 E. McDowell Rd.
Scottsdale, AZ 85252
Tel: (602) 441-7625
  "Peregrine DGPS–C-MAP Cartridges."

NAVIONICS
P.O. Box 722
Woods Hole, MA 02543
Tel: (800) 848-5896
Fax: (508) 548-9030
  Navigation-chart cartridges.

ROBERTSON MARINE ELECTRONICS
400 Oser Ave.
Hauppauge, NY 11788
Tel: (800) 645-3738
Fax: (516) 231-3178
  Electronic navigation-chart plotters.

SI-TEX
P.O. Box 6700
Clearwater, FL 34618
Tel: (813) 535-4681
Fax: (813) 530-7272
  "Nav-Add."

SIGNET MARINE
16321 Gothard #E
Huntington Beach, CA 92647
Tel: (714) 848-6467
Fax: (714) 848-6009
  "SmartMap."

SIMRAD
19210 33rd Ave. W.
Lynnwood, WA 98036
Tel: (206) 778-8821
Fax: (206) 771-7211
  "Simrad/Taiyo" color plotter.

Digitizers–Conventional Chart-to-Electronics Interface

CHART KIT/BETTER
BOATING ASSN.
P.O. Box 407
Needham, MA 02192
Tel: (800) 242-7854
Fax: (617) 449-0514
    "Yeoman."

KVH INDUSTRIES
110 Enterprise Ctr.
Middletown, RI 02840
Tel: (401) 847-3327
Fax: (401) 849-0045
    "Quadro," "Yeoman."

Radios and Radio-Navigation Interfaced Systems

Telex Communications Systems

HAL COMMUNICATIONS
P.O. Box 365
1201 W. Kenyon Rd.
Urbana, IL 61801
Tel: (217) 367-7373
Fax: (217) 367-1701
    Sitor/telex modems and systems.

HULL ELECTRONICS
1100-B No. Magnolia Ave.
El Cajon, CA 92020-1919
Tel: (619) 447-0036
Fax: (619) 444-0628
    Software for sending telex and fax over SSB radio.

RADIO HOLLAND U.S.A.
8943 Gulf Freeway
Houston, TX 77017
Tel: (713) 943-3325
Fax: (713) 943-3802
    "Sailor," "Thrane & Thrane."

SEA
7030 220th St. S.W.
Mountlake Terrace, WA 98043
Tel: (206) 771-2182
Fax: (206) 771-2650
    "Seator" radio/telex modem.

Navtex Systems

ALDEN ELECTRONICS
40 Washington St.
Westboro, MA 01581
Tel: (508) 366-8851
    Navtex, "Marinefax" weatherfax with Navtex.

FURUNO
271 Harbor Way
S. San Francisco, CA 94080

Manufacturers and Dealers   165

Tel: (415) 873-9393
Telex: 331-419WUI

RADIO HOLLAND U.S.A.
8943 Gulf Freeway
Houston, TX 77017
Tel: (713) 943-3325
Fax: (713) 943-3802
  "Lokata."

RAYTHEON MARINE
46 River Rd.
Hudson, NH 03051
Tel: (603) 881-5200
Fax: (603) 881-4756
  "JRC."

SHIPMATE
400 Oser Ave.
Suite 1400
Hauppauge, NY 11788
Tel: (516) 231-3000
Fax: (516) 231-3178

Weather Map Receivers and Hardware

ALDEN ELECTRONICS
40 Washington St.
Westboro, MA 01581
Tel: (508) 366-8851
  "Marinefax,"
"Faxmate" recorder.

FURUNO
271 Harbor Way
S. San Francisco, CA 94080
Tel: (415) 873-9393
Telex: 331-419WUI

PHITECHNOLOGIES
4605 N. Stiles
Oklahoma City, OK 73105
Tel: (405) 521-9000
Fax: (405) 524-4254
  "Nagrafax."

RAYTHEON MARINE
46 River Rd.
Hudson, NH 03051
Tel: (603) 881-5200
Fax: (603) 881-4756
  "JRC."

SEA
7030 220th St. S.W.
Mountlake Terrace, WA 98043
Tel: (206) 771-2182
Fax: (206) 771-2650
  "SeaFax" modem for connecting receiver to printer.

SIMRAD
19210 33rd Ave. W.
Lynnwood, WA 98036
Tel: (206) 778-8821
Fax: (206) 771-7211
  "Simrad/Taiyo."

SONY CORP. OF AMERICA
One Sony Dr.
Park Ridge, NJ 07656
Tel: (201) 930-6440
Fax: (201) 930-0491
  Portable radio with weatherfax.

UNIVERSAL SHORTWAVE
6830 American Pkwy.

Reynoldsburg, OH 43068
(614) 866-4267
Fax: (614) 866-2339
"Info-Tech" facsimile converter for SSB with computer printer..

GPS and GPS Computer Interface Equipment

APELCO MARINE ELECTRONICS
(A Raytheon Company)
46 River Rd.
Hudson, NH 03051
Tel: (603) 881-9605
Fax: (603) 881-4756

AUTOHELM
(A Raytheon Company)
46 River Rd.
Hudson, NH 03051
Tel: (603) 881-9605
Fax: (603) 881-4756
"ST-50."

BROOKES & GATEHOUSE
P.O. Box 308
New Whitfield St.
Guilford, CT 06437
Tel: (203) 453-4374
Fax: (203) 453-6109
"B&G," "Network," "Navstar."

CETREK WAGNER EAST
300 Oak St.
260 Corporate Pk.
Pembroke, MA 02359
Tel: (617) 826-7497

Fax: (617) 826-2495
"Cestar."

EURO MARINE TRADING
64 Halsey St. #27
Newport, RI 02840
Tel: (800) 222-7712
Fax: (401) 849-3230
"NKE Topline."

FURUNO
271 Harbor Way
S. San Francisco, CA 94080
Tel: (415) 873-9393
Telex: 331-419WUI

GARMIN
11206 Thompson Ave.
Lenexa, KN 66219
Tel: (913) 599-1515
Fax: (913) 599-2103

HUMMINBIRD
3 Humminbird Lane
Eufaula, AL 36027
Tel: (205) 687-6613
Fax: (205) 687-4272

ICOM
2380 116th Ave. N.E.

Manufacturers and Dealers 167

Bellevue, WA 98004
Tel: (206) 454-8155

INTERPHASE
TECHNOLOGIES INC.
1201 Shaffer Rd.
Santa Cruz, CA 95060
Tel: (408) 426-2007
Fax: (408) 426-0965
  "Star Pilot."

RAY JEFFERSON
4200 Mitchell St.
Philadelphia, PA 19128
Tel: (215) 487-2800

KODEN
INTERNATIONAL
77 Accord Park Dr.
Norwell, MA 02061
Tel: (617) 871-6223
Fax: (617) 871-6226

KVH INDUSTRIES
110 Enterprise Ctr.
Middletown, RI 02840
Tel: (401) 847-3327
Fax: (617) 849-0045
  "Quadro."

LORAN-CARD
2555 N. Dixie Highway
Lakeforth, FL 33460
Tel: (407) 547-8401
  "The GPS Card."

MAGELLAN SYSTEMS
960 Overland Court
San Dimas, CA 91773
Tel: (714) 394-500
Fax: (714) 394-7050

MARINETEK
2076 Zanker Rd.
San Jose, CA 95131
Tel: (408) 441-1661

MICROLOGIC
9610 DeSoto Ave.
Chatsworth, CA 91311
Tel: (818) 998-1216
Fax: (818) 709-3658
  "Explorer,"
"Supersport," "Admiral."

MOTOROLA
4000 Commercial Ave.
Northbrook, IL 60062
Tel: (708) 576-2828
Fax: (708) 576-5891
  "Traxar" hand-held GPS
systems.

MOTOROLA GEG
8201 E. McDowell Rd.
Scottsdale, AZ 85252
Tel: (602) 441-7625
  Standard, differential and
hand-held Navstar (see also
Brookes & Gatehouse,
above).

NORTHSTAR
30 Sudbury Rd.
Acton, MA 01720
Tel: (508) 897-6600
Fax: (508) 897-7241

PANASONIC
2 Panasonic Way
7B-3
Secaucus, NJ 07094
Tel: (201) 392-6305

PRONAV
(see Garmin, above)

RAYTHEON MARINE
46 River Rd.
Hudson, NH 03051
Tel: (603) 881-5200
Fax: (603) 881-4756
"Raystar."

SHIPMATE
400 Oser Ave.
Suite 1400
Hauppauge, NY 11788
Tel: (516) 231-3000
Fax: (516) 231-3178
Standard and differential GPS.

SI-TEX
P.O. Box 6700
Clearwater, FL 34618
Tel: (813) 535-4681
Fax: (813) 530-7272

SIGNET MARINE
16321 Gothard #E
Huntington Beach, CA 92647
Tel: (714) 848-6467
Fax: (714) 848-6009
"SmartNav."

SONY CORP. OF AMERICA
One Sony Dr.
Park Ridge, NJ 07656
Tel: (201) 930-6440
Fax: (201) 930-0491
"Pyxis" hand-held GPS receiver.

TECOM INDUSTRIES
9324 Topanga Canyon Blvd.
Chatsworth, CA 91311
Tel: (818) 341-4010
"Telenav."

TRIMBLE NAVIGATION
645 N. Mary Ave.
Sunnyvale, CA 94086
Tel: (800) 874-6253
Fax: (408) 991-7781
"10X Navigator," "NavGraphic II" GPS/Plotter, "Nav-Trac" GPS with plotter, "TransPak."

VDO-YAZAKI
P.O. Box 2897
980 Brooke Rd.
Winchester, VA 22601
Tel: (703) 665-0100
Fax: (703) 662-2515

Transit Satellite Navigation Receivers

BROOKES & GATEHOUSE
P.O. Box 308
New Whitfield St.
Guilford, CT 06437
Tel: (203) 453-4374
Fax: (203) 453-6109
"B&G," "Horizon" GPS.

FURUNO
271 Harbor Way
S. San Francisco, CA 94080
Tel: (415) 873-9393
Telex: 331-419WUI

MAGNAVOX ELECTRONIC SYSTEMS CO.
2829 Maricopa St.
Torrance, CA 90503
Tel: (213) 618-1200

RADAR DEVICES
2955 Merced St.
San Leandro, CA 94577
Tel: (510) 483-1953
Fax: (510) 351-7413
"RDI," "StarTrac."

TREMETRICS, INC.
2215 Grand Ave. Pkwy.
Austin, TX 78728-3812
Tel: (800) 477-7290
Fax: (512) 251-1596
"Bridgestar," "Transtar."

Multi-System Navigation Receivers

ALDEN ELECTRONICS
40 Washington St.
Westboro, MA 01581
Tel: (508) 366-8851
 "Marinefax" Navtex/
weather facsimile system.

FURUNO
271 Harbor Way
S. San Francisco, CA 94080
Tel: (415) 873-9393
Telex: 331-419WUI
 Navtex/weather facsimile.

NORTHSTAR
30 Sudbury Rd.
Acton, MA 01720
Tel: (508) 897-6600
Fax: (508) 897-7241
 Loran and GPS systems.

SHIPMATE
400 Oser Ave.
Suite 1400
Hauppauge, NY 11788
Tel: (516) 231-3000
Fax: (516) 231-3178

TRIMBLE NAVIGATION
645 N. Mary Ave.
Sunnyvale, CA 94086
Tel: (800) 874-6253
Fax: (408) 991-7781
 "10X Navigator"
Loran/GPS, "NavGraphic"
GPS/ Loran plotter.

Glossary of Equipment Terms

Here is a list of the most common computer terms you are likely to encounter as you evaluate, purchase, install, and work with a personal-computer system.

access: To locate the desired data.

access time: The elapsed time between the instant when data are called for from a storage device and the instant when the delivery operation is completed.

ANSI (American National Standards Institute): An organization that develops and approves standards in many fields.

application program: Software designed for a specific purpose, such as accounts receivable, billing, or inventory control.

architecture: The organization and interconnection of computer-system components.

arithmetic-logic unit: The part of a computing system containing the circuitry that does the adding, subtracting, multiplying, dividing, and comparing.

ASCII (American National Standard Code for Information Interchange): A standard code used to exchange information among data-processing and communications systems.

auxiliary storage: A storage that supplements the primary internal storage of a computer, often referred to as secondary storage.

Autoexec.bat: A special MS-DOS file used to start up the MS-DOS operating system in a configuration consistent with the needs of the operator. *See also* Config.sys.

backup: Alternate programs or equipment used in case the original is incapacitated.

BASIC (Beginners All-purpose Symbolic Instruction Code): A high-level interactive programming language frequently used with personal computers and in timesharing environments.

baud: A unit for measuring data-transmission speed.

binary digit: Either of the characters 0 or 1, abbreviated ".

bit.binary number system: A number system with a base of 2.

bit: *See* binary digit.

broadband channels: Communications channels, such

as those made possible by the use of laser beams and microwaves, that can transmit data at high speeds.

buffer: An area of the computer's memory in which MS-DOS or other operating programs store data.

bulletin-board system (BBS): A remote computer facility designed to enable the user to call in via computer modem to establish a link between two computers so that the user may obtain information, electronic mail, and/or other services.

byte: A group of adjacent bits, usually eight, operated on as a unit.

cache: A very-high-speed storage device used to speed up access to large or complicated applications programs.

cathode ray tube (CRT): An electronic tube with a screen on which information may be displayed.

central processing unit (CPU): The component of a computer system with the circuitry to control the interpretation and execution of instructions. The CPU includes primary storage, arithmetic-logic, and control sections.

character: A number, letter, or symbol that an operator types from the keyboard or views on a computer screen.

chip: A thin wafer of silicon on which integrated electronic components are deposited.

clock speed: The operating speed of a computer processor. That speed at which a computer accomplishes tasks.

command: An input or program that tells the operating system how to carry out a specific task.

computer: An electronic-symbol manipulating system designed and organized automatically to accept and store input data, process them, and produce output results under the direction of a detailed, step-by-step, stored program of instructions.

computer network: A processing complex consisting of two or more interconnected computers.

Config.sys: A file used by the MS-DOS operating system to configure the computer at start-up to load applications software and otherwise instruct the machine to conform to the needs of the user. *See also* Autoexec.bat.

control key (Ctrl): A key on the computer's keyboard used to invoke commands when used in conjunction with one or more other keys.

Copy: A command used under MS-DOS and other operating systems to duplicate files from one directory or disk to another.

cursor: The shape viewed on the screen, usually blinking and/or lighted, that shows the operator where the next character typed will appear, or where data input to the screen will be placed.

data: Facts; the raw material of information.

data base: A stored collection of the libraries of data that are needed by organizations and individuals to meet their information-processing and -retrieval requirements.

data base management system (DBMS): The comprehensive software system that builds, maintains, and provides access to a data base.

data processing: One or more operations performed on data to achieve a desired objective.

Del (Delete): A command, under MS-DOS and other operating systems, that instructs the computer to

erase a character, file, or block of data.
- device: An item of hardware, such as a disk drive, modem, or printer, that performs a specific task or function.
- device errors: faults that occur when a hardware device fails to perform its function, or is inaccessible to or uncontrollable by the computer.
- diagnostics: Error messages printed by a computer to indicate system problems and improper program instructions.
- Dir (Directory): A command that instructs the computer to display the contents of a file or directory on screen.
- directory: A listing or table of contents of a file or disk.
- disk: A revolving platter on which data and programs are stored.
- diskette: A floppy disk; a low-cost magnetic medium used for I/O and secondary storage purposes.
- disk storage: Physical space or area on a computer disk drive where information is stored.
- dot pitch: The size, i.e., height and width, of a character as displayed on a computer screen or printed on the page.
- download: To move data from a remote computer.
- drive: The physical opening or bay in which a disk-drive device is mounted.
- editor: An applications program that enables the user to manipulate text or other data on a computer.
- edlin: A simple line editor used with the MS-DOS system to manipulate text in program files.
- electronic mail (e-mail): A general term to describe the transmission of messages by the use of computing

systems and telecommunications facilities.

error message(s): On-screen information about a fault condition in the computer or with software or hardware functions. See the operating manual for the computer, software, or other device indicated in the specific error message.

facsimile (fax) machine: A machine used to transmit pictures, text, maps, etc., between geographically separated points. An image is scanned at a transmitting point and duplicated at a receiving point.

file: A collection of related records treated as a unit.

floppy disk: A magnetic storage medium used for storing files and data, usually inserted in a disk drive and removable for storage.

format: To prepare a disk or floppy disk to operate under the computer's operating system, and to receive and store data.

hard copy: Printed or filmed output in humanly readable form.

hardware: Physical equipment such as electronic, magnetic, and mechanical devices. Contrast with software.

input/output (I/O): Pertaining to the techniques, media, and devices used to achieve human/machine communication.

interface: A shared boundary, e.g., the boundary between two systems or devices.

K (kilobyte): An abbreviation for a value equal to 2^{10} or 1,024 bytes.

microcomputer: The smallest category of computer, consisting of a microprocessor and associated storage and input/output elements.

microprocessor: The basic arithmetic, logic, and storage elements required for processing, generally on one or a few integrated-circuit chips.

modem: A device that modulates and demodulates signals transmitted over voice-grade communication facilities.

narrow bandwidth channels: Communications channels that can only transmit data at slow speeds, e.g., telegraph channels.

network: An interconnection of computer systems and/or peripheral devices at dispersed locations that exchange data as necessary to perform the functions of the network.

on-line: A term describing persons, equipment, or devices that are in direct communication with a CPU, usually one at a remote location that is connected with the user via telecommunications hardware and related software.

operating system: An organized collection of software that controls the overall operations of a computer.

parity check: A method of checking the accuracy of binary data after those data have been transferred to or from storage. The number of 1-bits in a binary character is controlled by the addition or deletion of a parity bit.

peripherals: The input/output devices and auxiliary storage units of a computer system.

plotter: A device that converts computer output to a graphic, hard-copy form.

print: A command that instructs the operating system to send selected data to the printer.

printer: A device used to produce humanly readable computer output. A wide variety of impact and non-impact printers are currently available.

program: (1) A plan to achieve a solution to a problem; (2) to design, write, and test one or more routines; (3) a set of sequenced instructions to cause a computer to perform particular operations.

RAM (random-access memory): A storage device structured so that the time required to retrieve data is not significantly affected by the physical location of the data.

record: A collection of related items of data treated as a unit.

ROM (read-only memory): Generally, a solid-state storage chip programmed at the time of its manufacture that cannot be reprogrammed by the computer user.

software: A set of programs, documents, procedures, and routines associated with the operation of a computer system. Contrast with hardware.

telecommunications: Transmission of data between computer systems and/or terminals in different locations.

voice-grade channels: Medium-speed data-transmission channels that use telephone communications facilities.

word processing: The use of computers to create, view, edit, store, retrieve, and print text material.

write-protect tab: A small tab or slot in a floppy disk to prevent the computer's read-write head from accessing and writing information onto the disk. The purpose is to protect files and data on the disk from accidental or deliberate alteration or erasure.

Index

AC (Alternating Current), 46-49
Accutide 1.2, 135-37
Accu-Weather, 78
America On-Line, 152
Ami Pro, 82
Amperage
    instrumentation, 95
    in surge protection, 45-46
Artificial Horizon, 129
AUTOEXEC.BAT, 101-02
Autohelm, 106, 107, 117-19
Autolink RT, 15, 16
Autopilot, 17, 117-19

Battery charger, 48
Baud, 56
BBS (bulletin board systems), 15, 75, 88, 116
Bison Instruments 486, 9
    Explorer Portable, 50
BIX, 152-53
Bluewater Yacht Manager, 87
Boatowner's Energy Planner, 96
Body Finder, 132
Booting, 100

Calculators, manufacturers of, 158
Calendar Creator Plus, 134
Cannon Bubblejet printer, 63, 64
CD-ROM, 65-69, 93
    software examples, 66-69
Celestial Navigation, 129-30
Cellular communications, vii, 58-61
Charting software, 138
    for IBM, 138-46
    for Macintosh, 146-48
    systems manufacturers, 162-63

Checkit, 30
CHKDSK, 96-97
Circuit tester, 99
Choosing a computer, 19-37
    peripherals, 55-70
Climatology graphs, 78
C-Link, 15, 59, 61, 115-17
Clock speed, 30, 34
Collision Avoidance System, by ARPA, 132
Communications, computer assisted, 14-17
Compact Disks. *See* CD-ROM
Compass Deviation Control, 133
CompuAdd, 41
CompuServe, 14, 73, 76, 89, 151, 153
Computer manufacturers, hand-held, 158, 161-62
Computerizing the helm, 105-24
Comsat, 59, 115-17
Comsat Mobile Communications, 114
CONFIG.SYS, 101-02
Corrosion, 6, 94
Costs, vii, 7, 8, 16, 107
    desktop vs. laptop, 23-24, 25-26
    GPS systems, 115
    modems, 56
    on-line services, 77, 151-56
    printers, 62
    satellite communications, 59
    software, 73, 84-85, 86
CPU (Central Processing Unit), vi, 4, 24
    installation of, 23, 41, 42-44
    Sea PC, 29-30
Crane, D. F. and Associates, 6,

179

27-33, 126, 129
Crosstalk, 73

Databases, on-line, 75-77
Datalocker, 88
Davis PC AstroNavigator
    Celestial Software, 130-33
DC (Direct Current), 47-49
Dead Reckoning, 131, 133
Dehumidifiers, 92-93
Dell, 41
Delphi, 14, 73, 75, 76, 77, 89,
    151, 153-54
Design of boat, 40
Desktop vs. laptop computers,
    20-26
    desktop systems, 22
DESQview, 30
DIALOG, 75-76
DIFAX, 78
Differential GPS (DGPS), 113
Digitizers, manufacturers of,
    164
Disk storage, 35-36
DMA light-list waypoints, 139,
    144
Dow Jones News Retrieval, 75,
    154-55
DRAMM chips, 36
Dr. DOS 6.0, 12
Durability, 5-6, 50-51

Electrical system, boat, 44-46,
    99
Equipment manufacturers and
    dealers, 157-170,
Excel 4.0, by Microsoft, 12, 13,
    86

Fastening components, 42-44
Fax modem, vii, 5, 7, 74, 106
    software, 73-75
Files, recovering lost, 100-01
Floppy disks, 35
Ford, Dr. Clyde, 135
Freeware, 73, 88-89
Fuses, 46, 99

Galaxy Inmarsat-C, 15, 59, 61,
    105-06, 114-15
Gateway, 41
GEnie, 73, 75, 155
GHA (Greenwich Hour Angle),
    131
Global Inmarsat Consortium,
    114
Global PC Navigator System,
    132-33
GPS (Global Positioning
    System), 15, 60, 105-06,
    107, 109-14, 139, 143, 145,
    147
    manufacturers, 149-50,
    166-68
Graphical User Interface (GUI),
    11, 72

Hand-held navigation calcula-
    tors, 161-62
Harbormaster, 137-38
Hard disks, 4, 7, 8, 35-36, 93,
    94
    booting problems, 100
    "crash," 98, 100
    maintenance, 96-98
Hardware basics, 3-9
Hayes software, 73
"Hayes Compatible"
    modems, 56
Helm, computerizing, 105-24
H.O. 249, 128, 133

IBM or IBM-compatible, 7, 22,
    24
    costs of, 24-25
Inmarsat, 115, 116
    Inmarsat-C equipment man-
    ufacturers, 119-24
Installation, of components, 24
    of computers on board, 39-
    53
Internet, 14, 77
Inventory software, 10
Inverter, AC-DC, 28, 46, 47-49
"Iron Mike," 117

Index

J. Henry Celestial Navigation, 126-28

Keyboards
  desktop vs. laptop, 24
  mounting methods, 44
  protecting, 50
  Sea PC, 31
Kodak Diconix printer, 63, 64

Lamps, 52-53
Laptops, vii, 4, 5, 6, 20, 25, 26, 34, 92. *See also* Notebook computers
  software for, 85
LCD displays, 33
LEXIS, 76
Lighting, 52-53
LOP (Lines of Position), 131
Loran, 15, 117, 139, 142, 143, 145-46, 147
Lotus 1-2-3, 13, 86

Macintosh, 7, 11, 12, 22, 24, 73, 85, 87
  costs of, 24-25
  hard disk repairs, 100-02
  laptops, 25, 34
  printers, 63, 64
  processor chips, 34
  Quadra, 34
  software errors, 102-04
  terminal software, 73
  Yacht management software, 88
Macintosh HyperNavigator, 128-29
Magazines, 13, 126
Magellan GPS receiver, 111, 112
Mail, electronic, 14, 116
Maintenance, of computers, 91-104
Maintenance software, 10
Marine-use computers, 6
  manufacturers of, 157-70
Maripress News Service, 116
Mark 3 Sextant, 129

Memory, 35-37
Micro Mariner charting software, 138-41
  surveying/scientific system, 141-43
Micronautics Corp., 133-34
Microsoft MS-DOS 6.0, 12. *See also* MS-DOS
Microsoft Word, 12
  for Macintosh, 74
  for Windows, 36, 74, 82, 83
Microsoft Works, 13, 84, 85
  for Windows, 85
Modems, 5, 7, 16, 55-61, 116
  cellular, 58, 59
  for faxing, 5, 7, 56-58
  internal vs. external, 57-58
  software for, 71-75
Monitors, vi, 4
  full-size, 23
  protecting, 49-52
  Sea PC, 30
  VGA, 6, 23
Mouse, 11
MS-DOS, 12, 72, 75, 84, 96-97
Multipurpose software, 84-85

National Severe Storms Forecast Center, 79
National Videotext, 73, 75
National Weather Service, 78, 79
Nautical Almanac, 128, 131, 133
Navigation, 7, 105-124
  software, 126-33
  using computers for, 125-48
Navigation charts, computerized, 10, 17, 118
  on CD-ROM, 66
  software for, 138-48
Navigation receiver manufacturers
  multi-system, 169-70
  transit satellite, 169
Navigational weather software manufacturers, 158-60
Navlink, 143-45

NEC, 66
Newton, Apple, 21
NEXIS, 76
NMEA 0183 standard, 117-18, 141, 142, 143, 145, 148
NOAA, 78, 134, 135-36, 137, 146
Norton Utilities, 96-98
    Disk Doctor, 100-01
    Speed Disk, 97
Notebook computers, 4, 5, 21, 32-33, 58-59. *See also* Laptops
Nylon straps, as fasteners, 43

Ocean Navigator magazine, 126
OCR (Optical Character Recognition), 57
On-line services, 75-77, 88-89, 151-56
O/S2 software, 11

PageMaker, 13, 84
Parallel port, 30-31
Passagemaker, 87-88
PATH statement, 102
PC Astro Navigator, 129-30
PC Computing, 13
PC Magazine, 13
PC Tools, 97-98
Peripherals, 55-69
Plotters, electronic, manufacturers of, 164-65
Portable computers, 5. *See also* Laptops, Notebook
Powerbook, Apple, 25, 34
Power faults, 95
Power requirements, 46-47, 95
PowerStar inverter, 47-48
Power supply, computer, 99-100
Printers, 61-65
    dot-matrix, 62, 63
    ink-jet, 62, 63
    laser, 13, 62
    protecting, 49-52
Processor chips, 4, 30
    clock speed, 7
    DX, 33
    Pentium, 34
    SX, 33
    68000, 34
    80286, 34
    80386, 33-34, 36
    80486, 33-34, 36
    80586, 34
    8088, 34
Procomm Plus, 73
Prodigy, 73, 75, 156
Professional Mariner, 48

QEMM memory manager, 30
QuarkXPress, 13
Quattro Pro, 86

RA (Right Ascension), 131
Racing Prognostications, 133
Rack-mount computers, 6
Radar, 107, 118
    summary charts, 78
Radio communications, 16
Radio Direction Fix, 133
Radio-navigation interfaced systems, 164-66
Radios, manufacturers of, 164-66
RAM (Random Access Memory), 8, 35-37, 97
Reliability, 40
Role of seagoing computers, 1-18

SA (Selective Availability) of GPS, 113
Safeguarding computers, 48-52
Sailing BBS, 89, 152, 156
Satellite communications, 58, 116
SatNav, 143, 147
SeaMail, 116
Sea PC, 6, 27-31
Serial port, 30-31
SES (Ship Earth Station), 116
SHA (Sidereal-Hour Angle), 131
SIGs (Special Interest Groups), 14
Sight Clustering, 133

SIMM chips, 36
Smartcomm, 73
Software, 9-13, 15, 81-89
  accounting. *See* Software, spreadsheets
  basic requirements, 12
  bundled, 82-83, 84-85
  database, 13, 85
  errors, 102-04
  freeware, 73, 88-89
  modem/terminal, 71-75
  public domain, 88-89
  shareware, 73, 88-89
  spreadsheets, 7, 12, 13, 85, 86
  word processing, 12-13, 82-84
  yacht-management, 86-88
Space for computer, in boat, 40
SPS (Standard Positioning Service), 112-13
SSB (Single Side Band) radios, 16
Sta-Lock nuts, 42
Star-Finding, 133
Statpower, 48
Straps, for fastening
  galvanized, 42
  nylon, 43
  stainless steel, 42
Sunrise-Sunset software, 133
Supra fax modem, 56, 57, 74
Surge protectors, 44-46, 95, 100
System 7, 11, 12, 73, 75, 96

Telecommunications
  services, 151-56
  software, 71-75
  utilities, 71-79
Telephones, 15
Telix, 73
Temperature, effects of, 93
Tide. 1; Tide. 2, 133-35
Tide prediction programs, 133-38
  for PC, 133-37
  for Macintosh, 137-38
  manufacturers of, 158-60

Tide tables, computerized, 10, 133-38
Toshiba laptops, 58-59
Trace Engineering, 48
Trackball, 31
Traditional Celestial Navigation with J. Henry, 128
Trimble Navigation, 59, 60, 114-15
Troubleshooting, 91-104
Turning off computers, when to, 94-95
Twilight Planning, 132, 133

Ultrathin notebook, 32-33
Unicom, 73
Upper-Air Analyses, 78

Viruses, 98
Voltage meter, 95
Voltage requirements, 28-29, 45, 95, 99
Voyage Plans, 133

Watch system, computerized, 17
Water damage, preventing, 49
Waypoint Management Software, 145-46
Weight considerations, 41
Weather depiction charts, 78
Weather, equipment manufacturers, 158-60
Weatherfax, 15, 31, 59, 78
Weather map receivers, manufacturers of, 165-66
Weather Prognoses, 78
Weatherproofing, 49-52
Windows, software, 11, 12, 30, 72, 102, 103
Winfax Pro, 74, 75
Word processing, 7. *See also* under Software
WordStar, 11, 83
WYSIWYG, 13, 82

Zenith Data Systems, 41
Zeos Pocket PC, 21, 41, 50